ENERGY FROM BIOCONVERSION

OF WASTE MATERIALS

ENERGY FROM BIOCONVERSION OF WASTE MATERIALS

Dorothy J. De Renzo

NOYES DATA CORPORATION
Park Ridge, New Jersey, U.S.A.
1977

Copyright © 1977 by Noyes Data Corporation
No part of this book may be reproduced in any form
without permission in writing from the Publisher.
Library of Congress Catalog Card Number: 77-71660
ISBN: 0-8155-0656-2
Printed in the United States

Published in the United States of America by
Noyes Data Corporation
Noyes Building, Park Ridge, New Jersey 07656

FOREWORD

The allocation of this book into both our *Energy Technology* and *Pollution Technology Review Series* appears to be singularly apt at the present time.

While the principles of anaerobic decomposition of organic matter have been known for a long time, and infrequent use of this type of gas generation has been made in the past, the simultaneous application to the solid waste disposal problem is a fortuitous development indeed. It would appear that the utilization of "energy crops" can ease the energy crisis without discomfort or sacrifice of any kind.

The technological information found here is based on government-sponsored studies (often performed by highly qualified personnel from industrial companies) and one U.S. patent. In the United States we are fortunate in receiving direct help from the numerous surveys, together with active research and development programs that are being supported by the federal government.

Advanced composition and production methods developed by Noyes Data are employed to bring our new durably bound books to you in a minimum of time. Special techniques are used to close the gap between "manuscript" and "completed book." Industrial technology is progressing so rapidly that time-honored, conventional typesetting, binding and shipping methods are no longer suitable. We have bypassed the delays in the conventional book publishing cycle and provide the user with an effective and convenient means of reviewing up-to-date information in depth.

The table of contents is organized in such a way as to serve as a subject index and provides easy access to the information contained in this book. Reference lists follow each chapter in order to provide the reader with easy access to further information on this timely topic. The Bibliography at the end of the book lists the highly important government reports and a source of their purchase.

CONTENTS AND SUBJECT INDEX

INTRODUCTION 1

SOURCES OF WASTE MATERIALS FOR BIOCONVERSION 2
   Potential for Methane Fermentation of Waste Materials 2
   Estimated Solid Wastes Available for Methane Production 4
   Availability of Organic Matter on Statewide Basis 5
      Agricultural Crops 6
      Forests 8
      Urban Wastes 8
      Agricultural Wastes 9
      Animal Wastes 10
      Industrial Wastes 10
      Cost Considerations 12
   References 13

BIOCONVERSION MECHANISMS 15
   Anaerobic Decomposition Process 15
   Terminal Anaerobic Dissimilation of Organic Molecules 17
   Anaerobic Mechanisms for the Degradation of Cellulose 19
      Bacteria 19
      Protozoa 27
      Cellulases 28
      Chemical Products 32
   Controlling Factors in Methane Fermentation 33
      Evaluation of Stimulation Potential of Various Stimulants—Part I 34
      Evaluation of Stimulation Potential of Supposed Stimulants—Part II 38
      Effect of Inert Surface on Acetate Utilization Rate 42
      Effect of Asbestos, Asbestos Extract, Iron and No Additive on Acetate
         Utilization Rate 43
      Effect of $CaCO_3$ Solids on Acetate Utilization Rate 44
      Effect of Volatile Acids Concentration on Acetate Utilization Rate 45
      Effect of Detention Time on Acetate Utilization Rate 45
      Effect of Mixing on Acetate Utilization Rate 46

      Assay of Trace Organics for Stimulation of Acetate Utilization Rate 47
      Effects of Surface Charge, Complexing Agent and Other Factors 48
      Effect of Frequent Temperature Variation on Methane Production 52
  References 66

BIOCONVERSION OF SOLID WASTE AND SEWAGE SLUDGE 68
  University of Illinois Studies 68
    Anaerobic Processing Costs 68
    Fermentor Residue Dewatering 73
  Dynatech Study 75
    Process Description 77
    Computer Model Results 79
  Pfeffer-Dynatech Anaerobic Digestion System 80
  Supporting Work for Dynatech 85
    University of Massachusetts 85
    Massachusetts Institute of Technology 86
  Other Digestion Studies 88
    University of Pennsylvania 88
    University of California 90
    Clemson University 93
    Addition of Coal to Sewage Sludge to Increase Methane Production 96
    Synergistic Production of Methane from Refuse and Sludge 98
  Management Aspects of Refuse Bioconversion 101
  References 103

URBAN TRASH METHANATION—POCE 105
  Feed Preparation 105
    Control of Digester Feed 105
    Current Feed Preparation Experiments 106
    Feed Preparation Options 109
  Anaerobic Digestion 113
    The Microbial Process 113
    Gas Production 116
    Variables Affecting Anaerobic Digestion 117
    Digester Design 128
  Sludge Dewatering 130
    Methods 130
    Supernatant 131
    Cake 131
  Gas Preparation 132
    Desired Gas Characteristics 132
    Gas Scrubbing Technology 132
  References 137

ANIMAL WASTE DIGESTION 138
  Oregon State University Conversion System Based on Thermal
  Discharges 138
    Animal Waste Problems 138
    Proposed Animal Waste Management System 140
  Bioconversion Studies at the University of California (Berkeley) 147
    Digestion as a Sole Disposal Process 148
    Digestion as a Component in a Photosynthetic Treatment System 149
  Department of Agriculture Study 151

Other Reports on Animal Waste Digesters Used for Energy............152
References...154

INDUSTRIAL WASTE TREATMENTS 157
Treatment of Petrochemical Industry Wastewaters 157
Rum Distillery Slops Treatment 158
   Status of Technology of Rum Distillery Slops Digestion.............161
   Process Flow Sheet and Design Criteria.............................162
   Economic Analysis 165
Utilization of Biogas in Sugar Industry 171
Treatment of Winery Wastes 171
References..172

METHANE FROM ENERGY CROPS..173
Studies on Algae at University of California (Berkeley)............173
   Digestion of Algae..174
   Integrated Conversion System......................................175
University of Pennsylvania and UARL Studies 177
   Materials and Methods...178
   Results of Single-Stage Digestion—UARL............................187
   Results of Single-Stage Digestion—University of Pennsylvania.......203
   Two Stage Digestion 212
   Mariculture Investigations..219
   Conclusions...220
References..221

BIBLIOGRAPHY 222

INTRODUCTION

The supply of natural gas in the United States is rapidly approaching depletion and the necessity for finding new sources of fuel gas is becoming urgent. One method of supplementing the gas supply is by anaerobically digesting organic materials to produce methane which is the major constituent of natural gas. The anaerobic digestion process offers a number of advantages over other forms of energy conversion:

(1) It produces a relatively clean product gas which can be readily upgraded to pipeline quality.
(2) It efficiently converts the carbon of organic matter to a usable fuel.
(3) Many types of organic matter are potential feedstocks.
(4) It operates at low temperature and ambient pressure.
(5) It is safe and reliable.
(6) There is existing large scale gas technology available which makes implementation of the process to solve immediate energy needs particularly attractive.

An obvious source of organic materials for the process is the solid waste generated daily in every-increasing quantities from urban, agricultural and industrial sectors. The bioconversion of waste materials to methane provides a potential solution not only to the energy problem but also to the solid waste disposal problem. It has been estimated that anaerobic digestion of solid wastes could supply up to 20% of the U.S. natural gas requirements. The major portion of this book is concerned with converting waste materials to methane.

Other sources of organic matter will have to be made available, however, if the bioconversion process is to be utilized to its full potential. The cultivation of terrestrial, marine and fresh-water plants solely for the purpose of providing organic feedstock for anaerobic digesters will become necessary. The propagation of heretofore undesirable vegetation, such as algae, water hyacinth, and kelp, as "energy" crops offers unusual opportunities for maximum utilization of natural resources to supply future energy needs. The last chapter of this book describes some of the work being done in evaluating these future energy crops.

SOURCES OF WASTE MATERIALS FOR BIOCONVERSION

The sources of material for this chapter were the following reports: PB 231 149, ORNL-5056 and AD/A 002 212. For a complete bibliography of these reports, see page 222.

POTENTIAL FOR METHANE FERMENTATION OF WASTE MATERIALS

According to Dr. Perry McCarty of Stanford University, who was the keynote speaker at the Bioconversion Energy Research Conference held at the University of Massachusetts in June 1973, anaerobic digestion for the conversion of waste materials to methane gas is being looked upon today as one possible means to offset the increasing shortage of natural gas. Digestion of sludges resulting from treatment of wastewaters from the population of the United States could result in the production of 70 billion cubic feet of methane per year, and from the anaerobic decomposition of municipal refuse, an additional 2,000 billion cubic feet per year is theoretically possible.

In addition, if all animal wastes and crop residues were fermented, the resulting methane production could be double the above and yield an equivalent of about 20% of the current natural gas consumption. The potential for methane fermentation would thus appear great.

Anaerobic digestion of human wastes has been practiced for well over 100 years, and the resulting methane gas was used for practical purposes before the turn of the century. By 1963 anaerobic sludge treatment was practiced at over 70% of the wastewater treatment plants in the United States, serving over 60 million people. Thus, anaerobic treatment has been well implemented for municipal waste. However, for engineers, regulatory authorities, and treatment plant operators, anaerobic treatment is falling into disfavor. The trend is toward high-energy-consuming processes for treatment, including aerobic digestion and incineration. The reasons for this change in attitude need to be understood, evaluated, and overcome if methane fermentation is to have a significant impact in offsetting the energy crisis.

A major reason for decreasing emphasis on anaerobic treatment is a perceived lack of reliability in the process. A second significant reason is the increasing cost and difficulty in disposing of the solid residue resulting from anaerobic treatment, a problem for which incineration of the total sludge often becomes the most economical alternative.

There has been concern over lack of process reliability in anaerobic digestion for many years. A decreasing tolerance of foul odors emanating from upset digesters, stricter environmental requirements, and a willingness of the public to pay higher costs for environmental protection have all contributed to a search for more reliable processes for sludge disposal. Unfortunately, these alternative processes also require more energy. Thus, a significant improvement in process reliability is essential if methane fermentation is to compete with other processes.

Digester failure may result from natural microbial population imbalances, presence of inhibitory materials in wastes, operator error, or inadequacies in facility design. Natural population imbalances do not seem significant based upon long-term studies with readily digestible materials. Most failures have been blamed upon the presence of inhibitory materials, especially those entering the sewerage system from industry. Heavy metals receive the majority of the blame, perhaps mainly because they are always present rather than from good evidence that they are in fact the cause. There are indications that organic inhibitors may frequently be a cause, but proof has been difficult to obtain.

Failures due to error in operator judgement are difficult to document, but lack of knowledge of proper methods to prevent impending failures, inability to detect digester stress early, and failure to apply adequate control procedures in a prompt and appropriate fashion have no doubt resulted in many cases of complete inhibition of the process. Operators are greatly handicapped by the lack of a rapid and easy method for assessing the cause of digester imbalance. Development of a suitable procedure should be a high priority item. Inhibitory materials can be banned from the sewerage system but this can be effective only if a good monitoring procedure is available, another high priority item for development.

The last major factor resulting in poor reliability is inadequacies in design. The trend over the past twenty years has been to design digesters for shorter detention times since studies have shown efficient digestion is possible under such conditions. However, the factor of safety in many cases has been reduced to a point where any slight imbalance, unusual load, error in judgement, change in temperature, or other perturbation will result in failure of the system. More conservative design, inclusion of back-up units, monitoring devices, and automatic controls are features which may improve reliability. The need is for design criteria which will insure a system as reliable as others being proposed so that decisions about alternatives can be made in an objective fashion.

Essential to the survival of the anaerobic digestion process is the development of rapid and easy methods for the early detection of imminent digester failure, so that corrective measures can be taken to prevent it.

In relation to this problem, one method of monitoring digesters is a bioassay using microorganisms, instead of a chemical analysis. In a laboratory digester containing sludge from the digester, a mixed culture of methane bacteria and volatile fatty acids, is assayed for methane production. If methane is produced the problem is concluded to be an operational one. If no methane is produced, toxic materials are implicated.

There are various means of eliminating toxic materials: heavy metals by adding sulfide, LAS by adding long chain amines, H_2S by adding iron, and potential toxic organic compounds by extraction with a nontoxic substance such as methane.

The kinetics of degradation of the constituents of waste—cellulose, protein, and lipids—seem to be the following. Cellulose and protein are shown to be degraded rapidly at 37°C; however, methane fermentation requires five days. At 20°C, cellulose and protein are still degraded rapidly, but 12 days are required for methane fermentation. At 15°C, cellulose and protein are degraded much more slowly and methane fermentation requires 60 days. Lipids are found to undergo the slowest digestion in all three cases. Longer solids retention times result in higher efficiency. At short retention times microorganisms are washed out of the system faster than they can reproduce.

Considering all factors, methane fermentation has potential for the future only if increased reliability of the anaerobic digestion process can be economically achieved, and if methane fermentation can compete satisfactorily with other processes for obtaining energy contained in refuse.

ESTIMATED SOLID WASTES AVAILABLE FOR METHANE PRODUCTION

Cellulose is the world's most plentiful renewable resource. It comprises approximately one-half to one-third of all vegetation. Although large quantities are consumed as paper, lumber, textiles, and as feed for ruminants, most cellulose is unused and undergoes natural decay or is quickly discarded as domestic, industrial, or agricultural wastes. Reese, Mandels, and Weiss (1) indicate that roughly 0.05% of the total annual solar energy incident on the earth, or 5×10^{17} Btu, is made into cellulose. If this cellulose could be recovered, the energy content in biological products would be equivalent to roughly 5×10^{14} cubic feet of natural gas, worth roughly 500 billion to one trillion dollars per year.

Many of the cellulosic materials in which solar energy is stored are either widely scattered or inaccessible. These materials may be costly to collect, or in a form unsuitable for use. However, some cellulosic materials, in the form of domestic or industrial wastes, are available near cities or manufacturing areas. For other types of wastes, such as agricultural or food wastes, new techniques for collection are being developed. Although estimates of solid-waste production from various sources vary widely, they can be used to approximate the energy potential from various waste sources. The estimates used in Table 1.1 have been derived from the work of Anderson (2).

Urban refuse, crop residues, and manures are the major sources of organic material. However, of this group, only urban refuse is currently collected on a routine basis. Some crop residues are being collected, since equipment for stover collection is becoming available. Residues from the preparation of foods are also often available, though they are often produced on a seasonal basis. Because of

Sources of Waste Materials for Bioconversion

collection costs, which can account for about 80% of total disposal cost, urban wastes are routinely disposed near urban areas (3). On the other hand, crop and feedlot residues may be remote from energy-consuming centers, or from manufacturing plants which might use the waste cellulose for a feedstock. However, federal legislation requiring the provision of pollution control equipment for animal feedlot facilities and the availability of new farm machinery specific for the collection of crop residues may alter the situation (Federal Water Pollution Control Act, PL 92-500) (4)-(10).

Table 1.1 indicates both the amount of these waste materials and the quantities of methane that could be produced from these wastes. Since neither the overall quantities of the waste nor the average ultimate analysis of these wastes are reliably known, the assumption that the organic solids composition was $C_5H_7O_2N$ is not believed to introduce a serious error into the calculations.

TABLE 1.1: ESTIMATED SOLID WASTES FOR METHANE PRODUCTION

Source	Estimated Quantities (10^6 tons/year)		Calculated Methane Production* (ft^3)
	Organic Dry Wt	As Received	
Urban refuse	130	260	2.3×10^{12}
Crop and food residues	390	550	6.6×10^{12}
Manures	200	1,500-2,000	3.6×10^{12}
Industrial wastes	44	110	0.8×10^{12}
Logging and wood manufacture	55	80**	0.2×10^{12}
Sewage solids	12	–	0.9×10^{12}
Miscellaneous wastes	50	–	–
Total	881	2,500-3,000	14.4×10^{12}
		=	14.4×10^{15} Btu

*Calculations based on organic solids as $C_5H_7O_2N$ (53% carbon by weight); 80% recovery of carbon as gas; and a gas composition of 70% methane and 30% CO_2 (46% methane by weight). Thus (2,000) (1/113) (5) (0.7) (359) (0.8) = 17,800 ft^3 CH_4 per ton of organic solids. (One cubic foot of methane = 10^3 Btu).
**Varies widely. Calculated as 25% moisture.

Source: ORNL-5056

The energy potential available for recovery from waste products is considerable. For example, it may be seen from Table 1.1 that the domestic urban refuse contains roughly 130 million tons of dry organic solids. If all urban refuse were used to provide methane, it could provide about 2.3×10^{15} Btu or about 3% of the total current US energy consumption. If all organic wastes were routinely collected, they could provide as much as 14×10^{15} Btu, or about 20% of total US energy consumption, if converted to methane.

AVAILABILITY OF ORGANIC MATTER ON STATEWIDE BASIS

All green plants and organic wastes represent a reserve supply of solar energy that has been stored by the process of photosynthesis. While the caloric value of this material is relatively low (5,000 to 7,000 Btu/lb), it can be converted by established processes into more conventional fuels.

The RAND Corporation has made an assessment of the possible significance of fuel production from organic materials on a statewide basis. This study investigates the amounts of organic material that might be made available for energy purposes in the State of California, its potential fuel value, and the estimated cost. Sources of organic material that are considered are crops grown specifically for energy, the natural forests, and wastes from the urban, agricultural, and industrial sectors.

Agricultural Crops

Agricultural crops and managed stands of trees can convert about 0.5 to 1% of the light energy falling on a given area into organic matter. Even at these low efficiencies, the amount of energy that can be stored annually in a forest or field of crops is tremendous. Usually, cultivated plants achieve greater productivities than natural stands of vegetation, due to the inputs by the farmer. However, productivities depend heavily on the type of plant, soil, climate, etc.

Some productivities and prices for crops grown in California are given in Table 1.2. These are mostly crops grown for livestock feed that have been selected because they have a high yield of total plant matter. While they may not be the optimum plants for energy purposes, they are illustrative of what can be achieved on a commercial scale with good farm management practices. Yields for field crops are generally somewhat higher in California than in the central U.S.; however, production costs also tend to be $0.03-0.04/lb higher in California due to the greater use of irrigation and fertilizers, and the high land values. It should be kept in mind that for an energy crop to be practical, the energy inputs by the grower must be kept to a minimum.

TABLE 1.2: YIELD AND COST OF SELECTED CALIFORNIA CROPS*

Crop	Average Yield of Marketable Product (tons/acre)	Production Cost ($/ton)	Yield of Dry Organic Matter** (tons/acre)	Cost of Dry Organic Matter ($/lb)
Alfalfa hay	7.0	28.50	5.6	0.017
Oat hay	3.5	25.66	3.0	0.015
Corn silage	25.0	7.45	7.5	0.012
Sorghums, forage (double cropped)	45.0	5.88	12.6	0.011
Sorghums, grain	2.5	52.00	6.9	0.009
Sudan grass, green	48.0	3.28	5.8	0.014

*Reference (11).
**Corrected for moisture, inorganic content, and waste. See Reference (12).

Source: AD/A-002 212

The prices given in Table 1.2 for the illustrative crops suggest that organic material from crops will cost around $0.01/lb. It can be expected that the overall cost of producing plants for energy purposes might be somewhat less than regular food crops, since separate treatment of the marketable portion, such as fruit or seed, would not be necessary, and marketing costs might be avoided altogether.

If crops are to be grown for the purpose of providing raw material for fuel conversion, large areas of suitable land must be available. Land is a valuable commodity in California, and in some areas, there are already several conflicting uses for the same parcel. However, there is a considerable amount of farm land in the state that is capable of growing crops but is not now being used for that purpose. The amount and uses of farm land in California at the last census are shown in Table 1.3.

In the table, "cropland" refers to land that is suitable for growing crops without additional improvements; "other unharvested cropland" includes crops that failed or were unprofitable to harvest, summer fallow, soil improvement crops, and idle land; "woodland" includes all woodlots, timber tracts, cutover and deforested land that has value for growing wood products.

TABLE 1.3: LAND IN FARMS IN CALIFORNIA — 1969*

Land Use	Million Acres
Cropland	
Harvested	7.65
Used for pasture only	1.84
Other unharvested	1.75
Woodland	2.04
Permanent pasture	21.14
Other land (buildings, roads, etc.)	1.30
Total land in farms	35.72

*Reference (13)

Source: AD/A 002 212

In examining the table for land that might be suitable for growing energy crops, it can be noted that there are 3.6 million acres of unharvested cropland, and 2 million acres of woodland which could probably grow field or tree crops with a fairly high yield. It should be emphasized that these woodlands are not the large commercial forests used by the lumber industry. A very large portion of farmland is in permanent pasture, but this land is generally less productive than cropland or woodland, and an extensive effort to improve the yield, as for energy crops, might use more energy than it produced. Although it is not known how much of the present farm land actually could be diverted from present uses, it will be assumed for the purposes of the following calculations that all unharvested cropland and woodland is potentially usable for energy crops.

The greatest amount of potential land for energy crops is located in the central valleys. As the population increases, some of this land will no doubt be used for urban developments, most of which probably will be near the present metropolitan areas. However, even if all of the counties in the San Francisco Bay and South Coast regions are excluded, there are still nearly 5 million acres of potential cropland. If this land were producing energy crops at a yield of 6 tons/acre-yr, there would be enough organic matter for 250×10^{12} Btu/yr of fuel (assuming

60% conversion efficiency). This is 11.5% of the state's annual natural gas consumption.

Forests

About 40% of the land in California is covered with natural forests, 19 million acres of which are of commercial quality and 24 million acres are noncommercial (14). It is expected that the commercial forests will continue to be logged as much as possible for much-needed lumber. However, the noncommercial forests might provide some raw material for energy purposes. If the net productivity of the noncommercial forest is 3 tons/acre-yr, then the annual amount of stored energy is about 10^{15} Btu/yr. By removing an amount equal to the net annual productivity each year, and moving progressively from one area to another in the forest, it would theoretically be possible to maintain a continuous operation.

It is recognized that much of the California forest is needed for watershed protection, wildlife sanctuaries, and recreation. Large-scale logging operations are not compatible with these uses, especially if clear cutting is involved. Furthermore, much of the noncommercial forest is under pressure to be developed into second home, recreational, or resort areas.

Considering the difficulties of access to the high mountain forests, the need for maintaining the integrity of the watershed, and other competitive uses of the land, it seems highly questionable whether any of the standing forest could be exploited for energy purposes in California. However, in some areas, trees might be the optimum crop to grow for energy in the manner discussed in the previous section. Yields of 3 to 6 tons/acre have been achieved on tree farms in the Southeast (15), and mechanical methods of cultivating and harvesting, similar to conventional agriculture, can be used if trees are planted in rows and harvested young. The practice of coppicing can take advantage of energy stored in the tree roots, and extend the interval between plantings. Furthermore, tree plantations would be acceptable esthetically, even near populous areas.

Urban Wastes

Refuse: The amount of refuse generated in urban areas is a nearly direct function of population size, although larger cities tend to have somewhat more commercial refuse per capita than do smaller cities. A survey of waste generation in California yielded the following average factors for urban areas (16):

	Lbs/Cap-Day
Residential waste (paper, garbage, cans, bottles, etc.)	2.5
Commercial waste (office and business trash)	3.0
Municipal waste (street refuse, demolition, tree trimmings)	1.3
Total	6.8

The composition of urban refuse varies considerably, but analysis shows that on the average, about half of the weight is moisture and inorganic material. The other half may be considered dry organic matter that has fuel potential.

The population of California in 1972 was 20,524,000, of which 91.6% resided in standard metropolitan areas (14). It is assumed that essentially all of the refuse is collected in these areas, giving the total collected dry organic waste as:

$$20{,}524{,}000 \times 0.916 \times 6.8 \text{ lb/cap-day} \times 0.5 \text{ (dry organic)}$$
$$\times 365 \text{ days} = 23.3 \times 10^9 \text{ lb/yr}$$

This may be compared with the total refuse received at California disposal sites in 1967 of 39 billion pounds gross (16). At present most of this waste is buried in landfill operations or burned. Los Angeles County contributes 39% of the total, which is more by far than any other county.

Sewage: In a similar manner, the amount of sewage solids generated in California may be calculated. On the average, 0.18 to 0.22 lb/cap-day are generated, of which about 75%, or 0.15 lb/cap-day is dry organic matter (17). It is again assumed that essentially all the sewage in standard metropolitan areas is collected, so that the total amount for the state is 1.03×10^9 lb/yr.

Agricultural Wastes

Crops: A large percentage of agricultural crops is never used for food or feed, but ends up as waste in the field or processing plant. Field and orchard wastes include leaves, stalks, stubble, prunings, and culls. The amount of waste varies with the type of crop, being the largest for vegetables, intermediate for fruit and nut crops, and least for feed and grain. These wastes are generated over nearly 8 million acres of crops that are harvested in California, and present a collection problem if they are to be used. At present most crop wastes are collected on the farm and burned, or in some cases, they can be plowed back into the soil or used for fodder.

In the following calculations, it is assumed: (1) that all of the residue from fruit and nut crops could be collected, because it consists primarily of tree prunings which cannot be plowed into the soil nor fed to livestock; (2) that half of the field crop residue could be collected, since it is relatively easy to rake uniformly planted fields; and (3) that no vegetable crop waste is collected because these crops usually are grown in smaller quantities, in scattered locations, and are harvested at various times throughout the year. Using these assumptions, the total dry organic waste from crops can be calculated as indicated in Table 1.4.

TABLE 1.4: ORGANIC WASTE FROM CROPS IN CALIFORNIA

Crop	Acres Harvested in 1971* (thousand acres)	Waste Production Factors Fresh Waste** (lb/acre)	Waste Production Factors Dry Organic*** (lb/acre)	Assumed Collectible Fraction	Total Collectible Dry Organic (billion lb/yr)
Fruits and nuts	1,352	4,000	2,000	1	2.7
Field crops	5,759	3,000	2,400	½	6.7
Vegetables	731	6,000	1,800	0	0
Total					9.4

*Reference (14).
**Reference (16).
***Rand Corporation estimate.

Source: AD/A 002 212

Animal Wastes

Manure is one of the greatest waste disposal problems at present, and is becoming more acute because a greater number of animals are being raised on concentrated feedlots. Present methods of manure disposal include spreading it on land, burying, burning, or simply collecting it in piles. None of these are environmentally acceptable, and conversion to fuel would help alleviate the disposal problem, as well as contributing to the energy supply.

The total amounts of animal wastes that might be collected from farm animals in California are given in Table 1.5. Quantities were calculated on the basis of average solid waste production factors per animal. These factors actually vary over a wide range depending on the size of the animal, composition of feed and amount of confinement. In the table, only those animals that are raised in large herds or flocks in relatively confined areas are considered, so that collection of all the wastes can be considered practical (and indeed necessary). Broiler chickens generally reach marketable weight at age 2.5 to 3 months, so the total number raised annually was divided by 4.5 to allow for the short life span, and a smaller waste production factor was used to account for their smaller average size.

TABLE 1.5: ANIMAL WASTE PRODUCTION IN CALIFORNIA

Animal	Population on Confined Lots* (thousands)	Waste Production (Dry Organic Matter)	
		(lb/animal-day)**	(billion lb/yr)
Dairy cattle	1,341	10.0	4.89
Beef cattle on feedlots	1,045	7.0	2.67
Hogs and pigs	162	0.90	0.053
Sheep on feedlots	102	0.46	0.017
Chickens, layers	40,243	0.05	0.735
Chickens, broilers	19,700***	0.02***	0.144
Turkeys	16,801	0.08	0.490
Total			9.0

*Reference (14).
**References (18) and (19).
***Adjusted for short life span and small average size.

Source: AD/A 002 212

Industrial Wastes

Only the lumber and food processing industries will be considered here, since chemical and manufacturing industries have a very low proportion of organic material in their waste products.

Lumber Industry: The debris remaining after logging operations includes deadwood, culls, logs, branches, and brush. At present this waste may be left where it falls, stacked and burned at the site, or chipped for faster decay. These disposal methods cause fire hazards or air pollution or both.

It is estimated that for each board foot of harvested logs, there are two pounds of waste debris (16), about half of which is dry organic matter. It is difficult to imagine how much of this debris could reasonably be collected for energy uses. However, transportation is already made available for removal of saw timber from the forest, and sawmills are established within reasonable distance of logging operations. Therefore, it is assumed that logging debris could similarly be removed from at least the larger operations to suitably situated fuel conversion plants.

California timber production in 1970 was 4,566 million board feet (14). 57% of this was from the northern counties of Humboldt, Siskiyou, Shasta, and Del Norte, with the rest being scattered over much of the state. If just the debris from these counties were collected, the total dry organic waste from this source would be:

$$(4{,}566 \times 10^6 \text{ bd ft})(2 \text{ lb waste/bd ft})(0.5 \text{ dry organic})$$
$$(0.57 \text{ collected}) = 2.6 \text{ billion lb/yr}$$

Sawmills and planing mills produce further waste from the harvested logs, at the rate of 2.45 lb/bd ft (16). Efforts are being made to use more of this material in by-products such as pressed logs and wood pulp. The highest by-product utilization reported was 55% in Shasta County, and it is probable that other areas could achieve at least that much. The remainder of sawmill waste is usually burned at the mill. Alternatively, it could be available for energy conversion giving the following total amount for the state:

$$(4{,}566 \times 10^6 \text{ bd ft})(2.45 \text{ lb waste/bd ft})(0.5 \text{ dry organic})$$
$$(0.45 \text{ availability}) = 2.5 \text{ billion lb/yr}$$

Food Processing: There were 2,127 tons of wastes produced in California food processing industries in 1967, including fruit, vegetable and meat residues. 36% came from the five major food growing counties in the Central Valley, while another 15% came from Los Angeles County alone (16). Two-thirds of this waste is generated in August, September, and October, that is, during the harvest season. Thus the wastes tend to be concentrated both in location and time, which could add to the efficiency of collection. It is estimated that 35% of the wastes are being used in by-products, such as animal feeds and particle board, while the remainder is spread on the land, or disposed of in landfill, dumps, or sewage. Considerable effort is being made to find sanitary methods of disposal, because the high rate of spoilage of food wastes often causes land and water pollution.

The moisture content of food processing wastes varies tremendously from about 90% for fresh greens to 10% for grain hulls. It will be assumed for purposes of calculation that 50% of the waste is dry organic material, and that 65% is not used in by-products. Then the total dry organic waste for the state is 1.4 billion lb/yr.

A summary is given in Table 1.6 of the organic wastes that could reasonably be collected in California. The amount of fuel that could be produced from these wastes was calculated assuming a 60% energy conversion efficiency, which is commonly achieved with methane bacterial fermentations. The total fuel that could be produced from these wastes is about 8% of California's annual natural gas consumption. It is noted that urban refuse and crop wastes are the largest contributors.

TABLE 1.6: FUEL POTENTIAL FROM ORGANIC WASTES IN CALIFORNIA

Source	Collectible Wastes (billion lb/yr)	Caloric Value (Btu/lb)	Potential Fuel Production* (10^{12} Btu/yr)
Urban			
Refuse	23.3	5,000	69.9
Sewage	1.0	5,000	3.0
Industrial			
Logging	2.6	7,200	11.2
Sawmill	2.5	7,200	10.8
Food processing	1.4	6,000	5.0
Agricultural			
Manure	9.0	5,000	27.0
Crop waste	9.4	7,200	40.6
Total			167.5

*Assuming an energy conversion efficiency of 60%.

Source: AD/A 002 212

Cost Considerations

As shown in the previous sections, the fuel potential from both organic wastes and energy crops in California could be about 420×10^{12} Btu/yr, or about 19% of the state's current gas consumption. While this does not appear to offer a major source of supply, it could be of considerable importance in reducing the amount of imports, the need for superport construction, or offshore drilling.

How much land could actually be devoted to energy crops is open to question. Converting all of the currently unharvested cropland and farm woodlands to energy purposes would mean a 64% increase in the harvested acreage in the state. This is not a small consideration in terms of capital, labor and management in the agricultural industry. The present size of farms in California is quite large, with over 80% of the acreage on farms of more than 1,000 acres, which would perhaps make the transition to large-scale energy crop production easier than in some other states. The wastes from conventional food and feed crops could also contribute to the raw material for fuel conversion while providing a means of waste reduction.

The cost of producing fuel from organic matter can be estimated, although it will depend to some extent on the distance that the raw material must be transported and whether it would have had to be collected and disposed of anyway. The following calculation allows no credit for waste disposal, since the costs vary considerably for different types of waste and their location. The cost of conversion of sewage solids to methane via anaerobic bacterial digestion is approximately $0.31/million Btu, based on construction costs in Los Angeles. This process requires about 220 pounds of dry organic matter/million Btu of fuel, making it about 60% efficient in energy conversion.

If it is assumed that the raw material is transported an average of 10 miles at $0.07/ton-mile, that organic wastes cost nothing to obtain, and that energy crops cost $0.01/lb to produce, then the total cost of converted fuel would be $0.39 per million Btu for converted wastes, and $2.61/million Btu for converted crops.

Other conversion methods appear to be more expensive, although other forms of fuel may be more desirable in some cases. The wholesale prices of natural gas sold to electric utility plants was recently $0.35/million Btu in California (20). The projected costs of imported LNG range from $0.60 to $0.90 and for gas from coal gasification from $0.80 to $1.00. Thus it appears that fuel from energy crops in California is not competitive on a cost basis yet, but fuel from organic waste may be competitive, even without credit for waste disposal.

The use of organic wastes for energy in California is somewhat less attractive in cost and amount produced than in the United States as a whole. This is the result of several opposing factors: (1) the relatively high cost of growing crops in California, although yield per acre is somewhat higher also; (2) the higher than average percentage of cropland that is harvested in California; and (3) the high population density of California, which uses more total energy and more energy per capita than other states, but also produces more urban wastes.

REFERENCES

(1) Reese, E.T., Mandels, M., and Weiss, A.H., "Cellulose as a Novel Energy Source," *Advances in Biochemical Engineering*, eds. T.K. Ghose, A. Fiechter, and N. Blakebrough, Berlin: Springer-Verlag, 1972, pp. 181-200.
(2) Anderson, L.L., *Energy Potential from Organic Wastes: A Review of the Quantities and Sources,* U.S. Bureau of Mines Report No. 8549, 1972.
(3) Glysson, E.A., Packard, J.R., and Barnes, C.H., "The Problem of Solid-Waste Disposal," *Ingenor,* 9, Ann Arbor: University of Michigan, 1972.
(4) Henderson, P.A., "Some Economic Considerations Regarding the Use of Corn Crop Residues," *Nebraska Crop Residue Symposium,* Lincoln, Nebraska: University of Nebraska, 1973, pp. B1-B19.
(5) Petritz, D., and Parsons, S., *Custom Rates and Prices for Big Package Haymaking in Indiana,* Cooperative Extension Service Report No. ID-90, West Lafayette, Indiana: Purdue University, 1973.
(6) Parsons, S., *Some Thoughts on Big Package Hay,* Agricultural Extension Service Report No. AED-105, West Lafayette, Indiana: Purdue University, 1973.
(7) Weeks, S.A., Petersen, G.M., and Owen, F.G., "Evaluation of Mechanized Handling of Loose Hay," Paper No. 71-644, American Society of Agricultural Engineers Winter Meeting, December 7-10, 1971, Chicago, Illinois.
(8) Bledsoe, B.L., et al., "A Comparison of the Harvesting and Handling Characteristics of Large Hay Packages with Those of Conventional Bales," Paper No. 73-1564, American Society of Agricultural Engineers Winter Meeting, December 11-14, 1973, Chicago, Illinois.
(9) Gill, W.E., "Big Package Forage Handling and Feeding," Paper No. 73-1563, American Society of Agricultural Engineers Winter Meeting, December 11-14, 1973, Chicago, Illinois.
(10) Renoll, E.L., Antony, W.B., Smith, L.A., and Stallings, J.L., Paper No. 72-646, American Society of Agricultural Engineers Winter Meeting, December 11-15, 1972, Chicago, Illinois.
(11) University of California, Agricultural Extension Service, *Sample Production Costs,* various counties.
(12) NAS-NRC Committee on Animal Nutrition, *Table of Feed Composition,* Publication No. 1232, 1964.
(13) U.S. Dept. of Commerce, Bureau of the Census, *Census of Agriculture, 1969,* Vol. 1, Part 48, Sec. 2: California County Summary, May 1972.
(14) State of California, Documents Section, *California Statistical Abstract, 1972,* Sacramento, 1973.

(15) Steinbeck, "Short Rotation Forestry in the U.S.," American Institute of Chemical Engineers Meeting, New Orleans, La., March 11-15, 1973.

(16) State of California, Dept. of Public Health, *California Solid Waste Planning Study,* Vol. 1, September, 1968.

(17) Gloyna, E.F., *Waste Stabilization Ponds,* World Health Organization, Geneva, 1971.

(18) Loehr, R.C., *Pollution Implications of Animal Wastes: A Forward Oriented Review,* U.S. Dept. of Interior, Federal Water Pollution Control Administration, July 1968.

(19) Anderson, L.L., *Energy Potential from Organic Wastes,* U.S. Dept. of Interior, Bureau of Mines, Information Circular No. 8549, 1972.

(20) Edison Electric Institute, *Statistical Yearbook of the Electric Utility Industry for 1971,* October 1972.

BIOCONVERSION MECHANISMS

The sources of the material in this chapter are the following reports: N75-25292, PB 231 149, ORNL-5056 and PB 216 556. For a complete bibliography, see page 222.

ANAEROBIC DECOMPOSITION PROCESS

The anaerobic decomposition of any complex organic substance is basically a two-stage process. The first stage consists of the breakdown of the complex organic materials by acid formation bacteria into organic acids with the production of CO_2. These organic acids in the second stage are acted on by bacteria known as methane formers to produce CH_4 and CO_2.

In the reduction of organic material by anaerobic digestion, the organic complexes are acted upon by a group of floccules and anaerobic bacteria known as acid formers. The organic fraction of waste such as mixed municipal refuse consists of proteins, carbohydrates and fats. This material undergoes the acid fermentation which converts 35% of the material into short-chain organic acids. This consists of 15% being converted into propionic acid and 20% being converted into acetic acid. The other 65% of the organic material is converted into alcohols, aldehydes and long-chain fatty acids. These percentages are not exact and depend on the composition of the waste.

The initial stage of methane production or the acid-fermentation is essentially a constant BOD (biological oxygen demand) stage because the organic molecules are only rearranged. Thus, in the general case, the energy production by this step is very low and the microorganism growth is also low. Since most of the energy produced is used by the bacteria for growth, there is minimal energy liberated from the system. The first stage does not stabilize the solid waste, but it is essential for the second stage by converting the organic material to a form usable by the methane-producing bacteria.

The second stage takes place when anaerobic methane-forming bacteria act upon the short-chain organic acids. In this stage, the short-chain organic acids undergo methane fermentation with CO_2 acting as a hydrogen acceptor and being reduced to CH_4. The methane formed, being insoluble in water, escapes from the system and can be used for a fuel. The production and loss of methane cause the stabilization of the organic material. Since methane has a high energy content, most of the energy of the system goes into methane gas and not into the production of large amounts of cell mass and solids. Each cubic meter of methane produced at standard temperature and pressure removes 2.9 kilograms of COD or BOD (1,000 cubic feet of CH_4 at standard temperature and pressure removes 178 pounds COD or BOD).

The methane-producing bacteria consist of several different groups. Each group has the ability to ferment only specific compounds. Therefore, the bacterial mixture in a methane-producing system should include a number of different groups. When considering retention times of solids, the rate of bacteria production becomes important. For periods of 10 to 15 days of retention time, the rate of reduction is limited by methane fermentation. For systems where the retention time is longer than 15 days, the rate limiting aspect is then the hydrolysis of organic solids.

The actual destruction of organic material is directly related to the production of methane. An equation has been developed to predict the theoretical quantity of methane from the chemical composition of the waste.

$$C_nH_aO_b + (n - a/4 - b/2)H_2O \longrightarrow (n/2 - a/8 + b/4)CO_2 + (n/2 + a/8 - b/4)CH_4$$

Since a typical empirical chemical formula for the organic portion of mixed municipal refuse (MMR) is given by:

$$C_{30}H_{48}O_{19}N_{0.5}S_{0.05}$$

a theoretical methane value can be determined:

$$C_{30}H_{48}O_{19} + (30 - 48/4 - 19/2)H_2O \longrightarrow$$
$$(30/2 - 48/8 + 19/4)CO_2 + (30/2 + 48/8 - 19/4)CH_4$$

Since this calculation is approximate, the results can be given as:

$$C_{30}H_{48}O_{19} + 8.5H_2O \longrightarrow 13.75CO_2 + 16.25CH_4$$

This means that if 1 metric ton of MMR is considered, with an assumed organic content of 50%, the amount of water needed for the reaction is 112 kilograms (246.6 pounds), and the methane fermentation will produce 210.5 kilograms (464 pounds) of CO_2 and 298.7 kilograms (658.5 pounds) of CH_4. Since the slurry must be approximately 60% water for the methane bacteria, the water added per ton, assuming 24% initial moisture content, is approximately 200 kilograms (440 pounds) per metric ton. Obviously the system will not in general convert all the organic material in MMR to methane. Not only does the methane production depend upon the composition of the waste, but on a number of other environmental conditions. These conditions are temperature, reducing environment pH, nutrients and nontoxic conditions.

One of the most important operational parameters is the temperature in the reaction vessel. Increasing temperature will increase the rate of reaction. There are, depending on the methane bacteria present, two optimum temperature ranges. Mesophilic bacteria produce methane in the temperature range from 30°C (86°F) to 37.5°C (100°F) while thermophilic bacteria produce methane in the temperature range from 49°C (120°F) to 51°C (124°F). The reaction rates are much higher for thermophilic processes, but energy in the form of heating must be introduced into the system in order to maintain these temperatures.

The introduction of even small amounts of oxygen into the system will change it from a reducing environment and destroy the methane formers since they are strict anaerobes. Like most organisms, methane bacteria prefer pH values in the range of around 7.0. The actual range of optimum pH values is 6.6 to 7.6, and below 6.2 the acid conditions are quite toxic to methane bacteria, even though the acid-forming bacteria will continue to grow. The C/N ratio values are similar to those for aerobic bacteria, and again sewage sludge can be used to provide nutrient and nitrogen enrichment. The final requirement is that the system is free from toxic materials either in the form of inorganic salts or toxic organic compounds. This is a major problem in the case of MMR.

TERMINAL ANAEROBIC DISSIMILATION OF ORGANIC MOLECULES

At the Bioconversion Energy Research Conference held at Massachusetts University, Amherst, in June 1973, Dr. Paul Smith of the Department of Microbiology at the University of Florida (Gainesville) described work done on the identification of microorganisms involved in methanogenesis as well as the metabolic pathway of degradation of complex organic substances to CO_2 and methane. The generally held concept of the pathway is one in which insoluble organics are hydrolyzed by microbial extracellular enzymes to soluble organics, which are then broken down by acid-producing bacteria to volatile acids, H_2 and CO_2, which are finally metabolized by methanogenic bacteria to methane and CO_2.

If this were the pathway, one should be able to demonstrate each step and should be able to isolate microorganisms capable of performing the various steps; and also if the scheme were correct one should be unable to identify in large amounts any substances intermediate between volatile acids and methane.

Smith's work was centered on evidence to support the last three steps in an alternate scheme for methanogenesis:

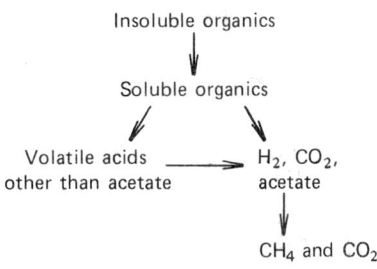

The important addition in the alternate scheme is the hydrogenogenesis from acids. The possible quantitative role of fatty acid methanogenesis was investigated using isotope dilution procedures. The results showed that fatty acids were precursors of most of the methane used. This was approached by labeling fatty acid pools to determine the extracellular pool size in the digestion process and to determine the turnover rate of labeled ^{14}C. Specific activity changed with time giving a rate constant which, when multiplied by the concentration, gave the rate of formation of the intermediate. This was correlated with the rate of methane formation.

Smith stated that for the procedure to be valid the reaction must be in steady state and the only substance contributing to the change in specific activity must be the formation of nonlabeled molecules from substrates. One can then determine the contribution of the intermediate to the overall methanogenic process. After four to eight hours the pool size was constant and the rate of methane formation was also constant. Samples from digesters were put in Warburgs in order to measure the rate of gas evolution (which was constant).

When 0.05 μmol of acetic acid labeled with ^{14}C was added in small amounts, so that pool sizes were not changed, the change in specific activity was found to be linear. A rate constant of 0.0052 per minute was obtained. With a pool size of acetate of 0.47 μmol per milliliter, the rate of formation of acetate was 0.024 μmol per milliliter per minute. Methane was evolved at a rate of 0.33 μmol of methane per minute. The contribution of acetate to the total methane produced was calculated to be 73%. Acetic acid was found to exist almost exclusively as extracellular pools; therefore, it was concluded that methane and acetate were produced by different organisms.

Smith did similar studies with labeled propionate. Similar kinetics were observed. With a rate constant of 0.0041 per minute and a pool size of 0.87 μmol per milliliter, the rate of formation of propionate was 0.0036 μmol per milliliter per minute. Since propionate was also found in the extracellular pool it was concluded that different organisms produce propionate and methane. It was determined that 13% of the total methane was formed from the reduction of CO_2 arising from the degradation of propionate. In addition propionate could also be converted to enough acetate to account for 17% more methane. Similar experimental procedures were used to show that butyrate contributes to 8% of the total methane by reduction of CO_2 and 32% by the conversion to acetate.

Smith next discussed the isolation of microorganisms that could perform the various steps in the conversion. Large numbers of organisms capable of metabolizing H_2 and CO_2 were isolated: a sarcina which could also metabolize methanol and acetate slowly; *Methanobacterium ruminantium;* a coccus; rods; and a spirillum, obtained from a propionate enrichment. No organisms were found which could use propionate or butyrate.

Lastly, studies of hydrogen utilization by domestic sludge were discussed. It was found that less methane was formed in the presence of H_2. Using labeled propionate and acetate, Smith found that H_2 inhibited propionate utilization, but had no effect on acetate turnover, which suggested that an important ecological role of the hydrogen oxidizing methanogenic bacteria in dissimilation processes is the maintenance of a hydrogen concentration low enough to prevent the inhibition of propionic acid metabolism and the concurrent cessation of the fermentation. Incubation of propionic acid enrichments under conditions which prevented formation of methane by CO_2 reduction resulted in the evolution of large quantities of molecular hydrogen. This was interpreted as showing that propionate metabolism is a hydrogenogenic process and not a methanogenic process. It was suggested that the same may be true for the metabolism of other fatty acids.

The data presented thus supported the concept that the digestion of complex organic substances consists of four stages: hydrolysis of complex organic molecules, acid production, hydrogenogenesis from acids and methane formation.

ANAEROBIC MECHANISMS FOR THE DEGRADATION OF CELLULOSE

If cellulosic materials are to be converted biologically to desirable products such as volatile acids or methane, it is necessary to degrade the long-chain cellulose molecule to smaller, soluble molecules which are composed of one to four glucoses. Fermentation can then be used to produce the desired chemicals from these smaller sugar molecules. Cellulose degradation can be accomplished by various species of anaerobic and facultative bacteria, protozoa and fungi (1). A description of the investigative work done in identifying these microorganisms follows.

The various fermentations reported have generally been performed only at benchscale. Most of the work has been done as either part of ruminant nutrition studies, or has been performed as research in anaerobic microbiology. Little emphasis has been given to the types of kinetic and strain selection studies which would be useful as a basis for the design of large-scale or pilot-scale process equipment. Although equipment capable of performing the aerobic fermentations which are used to produce various biochemicals is available, it is not known whether this equipment will be applicable to the fermentation of cellulose under strictly anaerobic conditions.

Bacteria

Because of the economic importance of cellulose degradation in the rumen, mixed cultures of organisms able to ferment cellulose were studied during the late nineteenth century. However, success in the culture of pure strains of rumen-derived cellulolytic bacteria was not attained until the nutritional requirements of these organisms were elucidated (1). During the 1940s, development of rumen-fluid media and hard-rubber stoppers permitted the isolation of several strains of cellulose-fermenting bacteria (2). To date, several different genera have been isolated from the rumen. Pure cultures of Bacteroides, Butyrivibrio, Clostridia, Cillobacterium, Cellulomonas and Ruminococcus strains are known to ferment cellulose.

In spite of the progress to date, the culturing of these organisms is still difficult. However, there is a considerable body of published reports from which to draw information on the behavior of anaerobic cellulose-fermenting bacteria.

Using direct microscopic examination of formalin-preserved rumen contents, Baker (3) was able to make a semiquantitative determination of the microorganisms in the rumen of freshly slaughtered oxen. He reported that certain iodophile microorganisms decompose cellulose in the rumen. This was inferred from the position of these microorganisms within cavities in cellulose fibers. Because the fibers observed were resistant to attack by 17.5% NaOH, Baker concluded that the microorganisms causing cavities in these fibers were capable of degrading α-cellulose. However, where fibers were either strongly lignified or contained cutin, they showed little effect. Baker reported that cellulose decomposition was due primarily to vibrios and to iodophilic cocci. He concluded that the decomposition of cellulose resulted in the liberation of sugars which were converted into iodine-staining polysaccharides in the cocci, and that the process of cellulose degradation was also accompanied by the production of various organic acids.

Using an acid-swollen cellulose-mineral salts-agar medium, Hungate (2) isolated an anaerobic cellulose-digesting bacterium, *Clostridium cellobioparus*. The organism is a slightly curved gram-negative spore-forming rod, roughly 3 to 5 μ by 0.3 to 0.4 μ. Biotin is required for growth, as is a strictly anaerobic environment. Glucose, cellulose, fructose, xylose, arabinose, mannose, cellobiose, melibiose, maltose and a birch-derived hemicellulose were readily fermented. Sucrose, lactose, raffinose, galactose, mannitol and dextrin were slowly fermented. No acid or gas formation was observed with melezitose, trehalose, rhamnose, salicin, inulin, glycerol or soluble starch.

Fermentation products from either glucose or cellobiose include hydrogen, carbon dioxide, acetic acid, formic acid and ethyl alcohol. The products from cellulose hydrolysis include cellobiose. Hydrolysis of cellulose continued after the culture ceased dividing and formed spores, indicating that the cellulase produced by the organisms is an extracellular product.

Hungate reported the isolation of an obligately anaerobic actinomycete from the gut of the termite, *Amitermes minimus*. The actinomycete, *Micromonospora propionici*, is capable of fermenting glucose or cellulose in a complex medium (1). The organism appeared to have an additional requirement for meat extract, yeast or dried grass in the medium. However, these materials could not be replaced by a B-vitamin mixture containing pantothenic acid, thiamine, nicotinic acid, riboflavin, pyridoxine, biotin and folic acid. Fermentation products appeared to be carbon dioxide, acetic acid and propionic acid. Fermentation products accounted for roughly 70% of the substrate carbon. The fermentation was, in many ways, similar to the fermentations of propionic acid bacteria.

Hungate reported that it was very difficult to obtain subcultures of rumen bacteria capable of fermenting cellulose when media using hay extract, grass extract or finely divided cellulose were used (4). The medium was modified to have a carbon dioxide-sodium bicarbonate buffer system, to contain boiled rumen fluid, and to use finely milled swollen cellulose. This modification permitted the rapid and simple subculture of cellulose-degrading bacteria. Three strains of cellulose-fermenting cocci and a strain of cellulose-degrading rod bacteria were isolated. The rod form is capable of migration through agar media. The cocci appear to be similar to those described by Baker (3). The absence of polysaccharide grains from the rod organism indicated that the storage of polysaccharides was not a definitive test for the degradation of cellulose.

Enebo (5) demonstrated the symbiosis of various anaerobic cultures which ferment a single substrate. He separated a culture of anaerobic thermophilic cellulose-digesting organisms into three strains. These were: (1) a strain which fermented cellulose to produce lactic, formic and acetic acids and ethanol; (2) a strain capable of producing lactic, formic, acetic and butyric acids and ethanol from dextrose but incapable of fermenting cellulose or dextrin; and (3) a strain capable of growing aerobically on nutrient medium whose fermentation products are similar to those of the cellulose-degrading strain. When the three cultures are grown together, butyric acid, rather than the lactic acid predominating pure cultures, dominates the products produced. Enebo indicates that this is due to a relationship between butyric and lactic acid production.

Hungate (6) reviewed the state of the art of cultivating anaerobic mesophilic cellulolytic bacteria. He described the method for the cultivation of these bacteria in considerable detail. The major nutritional requirements for those cellulolytic bacteria isolated from the rumen were pH control, rumen fluid-based media, and very finely divided swollen cotton cellulose. Strict anaerobic conditions were also required. Using this method, over 25 separate strains of cellulolytic bacteria were separated from rumen contents by Hungate and his graduate students.

Bacteroides succinogenes, which produces succinic and volatile acids from cellulose and from cellobiose, was isolated by serial transfer from rumen fluid. The major volatile acid formed appeared to be acetic acid. The organism does not form distinct colonies, but generally migrates to the edge of the cleared zone in cellulose agar. A rod organism capable of fermenting several sugars and cellulose was subsequently isolated. The organism was capable of fermenting arabinose, xylose, glucose, galactose, mannose, levulose, maltose, lactose, cellobiose, sucrose, raffinose, dextrin, soluble starch and hemicellulose with the production of acid and gas. It did not, however, completely ferment cellulose in the fashion of *Bacteroides succinogenes.*

This appears to be the first instance of an organism incompletely fermenting cellulose mentioned in U.S. literature, although partial cellulose-fermenting organisms were isolated earlier in Europe. In the rumen, the rod organism isolated by Hungate may ferment either hemicellulose or β-cellulose, rather than α-cellulose.

Hungate also isolated various strains of cellulolytic organisms from the Moscow, Idaho, sewage sludge digester. These bacteria were isolated only from anaerobic sludge, not from raw, aerobic sludge. Isolates were made using rumen fluid agar in the same manner that isolates of rumen cellulolytic bacteria were obtained. However, the isolates derived from sewage sludge proved to have nutritional requirements which could be supplied by the addition of small amounts (0.05%) of yeast extract to a mineral salts medium. The cellulolytic cultures derived from sewage sludge did, however, have one major difference from the cultures derived from rumen fluid: they carried the fermentation of cellulose to glucose, rather than to cellobiose or to volatile acids.

Isolates from soil samples were also prepared by Hungate. The fermentation products and the organisms appear to be similar to those isolated from sewage sludge. However, isolates from soil produced a slightly higher amount of volatile acids than did isolates from sewage sludge samples.

Colorless rumen cocci, which appear to be the most abundant organisms in the rumen, were isolated using rumen fluid medium. These organisms showed a considerable preference for cellulose or cellobiose as a substrate, but were, in some cases, unable to use glucose as a substrate. Some essential nutrient found in rumen fluid, but not found in grass, malt sprouts, nutrient broth or Gall's medium was required. The missing nutrient could, to a slight extent, be found in, or replaced by, either proteose peptone or distiller's dried solubles. However, only rumen fluid supported good growth. Differences in the proportion of fermentation products were noted when cellobiose, rather than cellulose, was used as a substrate. Cellobiose cultures produced more lactic acid, but less hydrogen, carbon dioxide and ethanol, than did cellulose cultures. Neither propionic or butyric acid were formed in any of the cultures.

Some yellow-pigmented rumen cocci were also isolated. These organisms appeared to have a fermentation similar to the colorless rumen cocci. However, the organisms do not appear to ferment cellobiose or glucose, but could only be isolated using cellulose. The cultures were difficult to transfer, and most of the isolates died.

Hungate reports the isolation of various types of spore-forming cellulolytic rods. He reports that these organisms made up the bulk of those isolated from enrichment cultures by earlier investigators. However, these organisms, though fairly easy to isolate from rumen enrichment cultures, are not numerically important in the rumen. Cellulose and hemicellulose seem to be favored substrates for the organisms. However, a minimal mineral-cellulose medium would not support growth. Either rumen fluid, or a medium supplement derived from whole living material, was required. The main fermentation products from cellulosic spore-forming rods include carbon dioxide, hydrogen, ethyl alcohol and acetic acids. Other volatile acids and succinic or lactic acid may also be formed.

McBee (7) discussed early attempts at isolating thermophilic cellulose-digesting organisms. Most of these early attempts resulted in cultures which behaved erratically, or which appeared not to be pure. However, McBee, using the techniques developed by Hungate (6), was successful in the isolation of some strains of thermophilic cellulolytic organisms. Various attempts were made during the 1940s to develop a stable industrial cellulose fermentation using thermophilic cellulose-digesting organisms. However, there was considerable difficulty in product recovery and cellulose preparation. Very small quantities of cellulose were able, however, to be fermented. Generally, the proportion of cellulose which could be fermented anaerobically by thermophilic microorganisms is as low as 0.2%. This makes product recovery somewhat difficult.

The fermentation products which are produced by thermophilic cellulose-digesting organisms generally appear to consist of at least acetic acid, carbon dioxide and hydrogen. In some studies, other volatile acids, and lactic acid, have been observed as culture products. Most of the cellulosic thermophiles appear to be slender, spore-forming gram-negative rods. They have been isolated from several different types of enrichment cultures, including those derived from soil, river mud, saline mud and horse manure. Only some cultures are motile. Although some cultures were isolated from environments conducive to having a number of nutritional requirements, all of the isolates compared by McBee were able to grow on a cellulose-mineral-salts medium to which 0.05% yeast extract was added. A medium of defined composition could be used if several B vitamins, including thiamine, riboflavin, pantothenate, pyridoxine and biotin were added to the mineral-salts medium listed above. Growth was vigorous, and daily transfer of cultures was performed. All of the cultures tested preferred a temperature range of 55° to 65°C; growth was limited below 50°C. All of the cultures tested by McBee were able to ferment cellulose, cellobiose, xylose and hemicellulose. These cultures were not able to ferment glucose, fructose, mannose, galactose, arabinose, sucrose, lactose, maltose, melibiose, trehalose, inulin, salicin, dextrin, soluble starch, inositol, sorbitol, dulcitol, glycerol, pectin and gum arabic. Growth of the organisms tested only occurred between pH 6.4 and 7.4. Ammonium salts were found to be an adequate source of nitrogen for growth. Of the six cultures tested, McBee found that only one was able to ferment essentially all of the cellulose supplied. Other cultures were able to ferment roughly 50 to 60% of the cellulose supplied. Products of all the cultures were as described above. McBee recommended that all of the cultures tested be considered strains of the same organism.

Hammerstrom, Claus, Coghlan and McBee (8) determined that the cellulases of *Clostridium thermocellum* and some Cellulomonas species are constitutive. They determined this by culturing the species on several noncellulose substrates. Some cellulolytic activity was produced by all of the cultures tested. However, the activity did vary somewhat with the substrate tested. They did not make a determination of whether this change in activity with substrate was due to failure of the organism to grow well, or due to the induction or repression of some of the organisms' cellulase enzymes by products of cellulose hydrolysis. Enzyme activity was measured as either reduction in the viscosity of CMC or by production of reducing sugar by culture filtrates of the Cellulomonas or Clostridium species.

Kitts and Underkofler (9) determined that rumen fermentation could be arrested at the reducing sugar stage by the addition of toluene, thymol or sodium fluoride. Cultures so treated produced primarily glucose, which Kitts and Underkofler presume to be the main intermediate of rumen fermentations. They found that preparations of cellulase could best be made from the microorganisms themselves, rather than from the rumen fluid supernatant. Paper chromatography was used to determine the amount of product formed. It is possible that the analytical method used to detect glucose was in error, since no cellobiose was found. This directly contradicts the findings of Hungate (6). There is also a possibility that the cellulosic organism isolated during these experiments did not arise from the rumen inoculum, but was introduced into the rumen with the cow's feed.

Halliwell (10) investigated the degradation of cellulose and cellulose derivatives with both *Myrothecium verrucaria*, extracts from rumen cultures, and whole rumen enrichment cultures. He determined that cellulolysis required at least two enzymes, one of which attacked the cellulose so as to make it more readily degradable, and the other which performed the actual degradation. Mixed cultures were used for the work because of the inability of the pure cultures currently available to completely degrade cellulose.

An enzyme extract of rumen enrichment cultures was prepared by grinding the rumen enrichment culture with alumina powder in a mortar and subsequently extracting the alumina with phosphate/sulfide buffer, pH 7.2. This extract possessed appreciable activity against Whatman cellulose fibers, being able to solubilize up to 30% of these fibers in 72 hours. A butanol extract of rumen enrichment cultures, although possessing only the ability to solubilize 10% of the total weight of cellulose fibers, was able to decrease the viscosity of carboxymethylcellulose appreciably. A freeze-dried extract of whole rumen contents was able to solubilize up to 80% of the total weight of Whatman cellulose fibers within 72 hours.

Halliwell noted that the cellulases of rumen microorganisms are very unstable, and must be handled anaerobically at low temperatures. Filtrates from *Myrothecium verrucaria* extracts growing on cellulose were also prepared. Their behavior appeared to be similar to that of the alumina-ground rumen extract when tested on Whatman cellulose, but did appear to be somewhat higher when tested on swollen cellulose powder. The *Myrothecium verrucaria* extracts appeared to be able to solubilize roughly 88% of swollen cellulose powder. This value is considerably above the values which have been determined by other investigators.

Halliwell makes the suggestion that some of the results which have been reported by others investigating the cellulase action on relatively unaltered natural cellulose are caused by the hydrolysis of xylans or glucosans which are naturally associated with cellulose. He contends further that the usual reducing sugar tests which are employed to quantify cellulase activity may thereby yield erroneous information.

In addition, Halliwell points out that the observed action of the cellulase enzymes in a given preparation may vary with the temperature of incubation. For example, the specificity of *Myrothecium verrucaria* preparations, as well as their relative activities on carboxymethylcellulose, changes drastically with temperature. Halliwell further indicates that no preparations are available which approach the vigor and activity of the live rumen culture in degrading cellulose.

In vitro studies were conducted by Goto and Okabe (11) with 13 species of Pseudomonas, 7 species of Xanthomonas, 2 species of Erwinia, 1 species of Agrobacterium and 1 species of Corynebacterium, including several isolates of each species to determine whether cellulolytic enzymes are produced typically by phytophathogenic bacteria.

It can be inferred from the results obtained from potato-decoction gel cultures that cellulolytic enzymes are produced from *P. setariae, P. solanacearum, X. campestris, X. citri, X. nigromaculans, X. oryzae, X. pisi* (n. sp.), *X. pruni, X. vesicotorio, E. carotovora, E. milletine, C. sepedonicum*, but not from *P. aeroginosa, P. andropogoni, P. cichorii, P. eriobotryae, P. iridicola, P. maculicola, P. marginalis, P. phaseolicola, P. syringae, P. tabali, P. theae* and *A. tumefaciens*.

The fermentation rates in the caecum, large intestine, and rumen of a number of mammals was studied in Kenya by Hungate (12). Characteristic clearing of the cellulose agar resembling that due to *Bacteroides succinogenes* from cattle (6) was observed in cultures from the eland, kongoni, zebu and camel. A few colonies resembling Butyrivibrio were also seen. In culture series where both noncellulolytic and total colony counts could be made, the counts were similar to those obtained with cattle by comparable methods.

Husain and Kelman (13) investigated the action of *Pseudomonas solanacearum* culture filtrates on cuttings of several plants. The action by these filtrates and by *Myrothecium verrucaria* cellulases was compared. The production of three different enzymes by the Pseudomonas culture was demonstrated. These enzymes, which are pectin methyl esterase, polygalacturonase and a cellulase, are believed to be responsible for the organisms' pathogenicity. These enzymes break down the host activity and cause a wilting effect that can, to a certain extent, be duplicated by the culture filtrates and by commercial enzyme preparations known to have equivalent functions.

Some of the nutritional requirements of the genus Ruminococcus have been studied by Bryant and Robinson (14)(15). Hungate (16) reviewed the ecology of rumen microflora and microfauna. He estimated that most of the highly represented rumen species, or species which fill these niches in rumen ecology, have been cultured. However, he also indicated that there is considerable discrepancy between the culture counts, as determined by colony counts, and the number of microorganisms determined directly by microscope counts of whole rumen samples.

Hungate states that *Bacteroides succinogenes, Butyrivibrio fibrisolvens,* Ruminococcus species and *Clostridium lochheadii* are among the major cellulose-decomposing organisms in the rumen. As previously mentioned, these organisms all produce carbon dioxide, hydrogen, acetate and formate, and most of them produce volatile acids, lactate or succinate from cellulose.

A method for routine counts of rumen bacteria capable of fermenting cellulose has been described in detail by Kistner (17). Storvick and King (18) investigated the cellulase enzymes produced by *Cellvibrio gilvus.* They determined that the organism produced several cellulase components, and that the cellulases produced cellobiose from cellulose. The preparation did, however, show a very low activity, appeared to be incomplete in enzyme activators and to contain an inhibitor.

Moscatelli, Ham and Rickes (19) report the determination of a β-glucanase in enzyme preparations derived from *Bacillus subtilis.* Strider and Winstead (20) reported the production of enzymes, including a cellulase, capable of dissolving cucumber cell walls, by *Cladosporium cucumerinum.* Presence of a cellulase, or cellulases, in filtrates from cultures grown using carboxymethylcellulose, sodium pectinol and sucrose as a carbon source was demonstrated. Cellulase activity was heat labile.

Brooks (21) reviewed the microorganisms indigenous to healthy insects. She indicated that only those insects which feed either on a highly polymerized and degradation-resistant material, or those which feed on a material which is missing certain necessary insect nutrients, harbor indegenous symbiotic microorganisms. These microorganisms serve to degrade insoluble polymers or to provide the necessary nutrients for the host insect. In the case of termites, much of the degradation of wood is accomplished by the use of various anaerobic bacteria and protozoa. The products of these organisms are generally assumed to be acetic acid, carbon dioxide and hydrogen. This has been demonstrated by some laboratory experiments (22).

Hungate (23) reviewed the state of the art in rumen bacterial culture. He stated that the following bacteria were important in the digestion of cellulose in the rumen: *Bacteroides succinogenes, Ruminococcus flavefaciens, Ruminococcus albus,* Butyrivibrios, Clostridia and *Cillobacterium ruminantium.* He stated that *Clostridium lochheadii* appears to produce large amounts of an extracellular cellulase. He also determined that cellulose-digesting bacteria, by count, appear to represent only 1 to 5% of the total rumen microflora. However, they perform one-third of the total degradation upon which the whole microflora depends for its nutrition. There is a major difference between whole rumen cultures and pure laboratory strains in that whole rumen cultures are capable of degrading cellulose substrates, such as cotton or whole filter paper, which are not appreciably attacked by the pure cultures currently being tested.

An apparatus for the continuous bench-scale culture of strict anaerobes was described by Hobson and Smith (24). A continuous flow of purified carbon dioxide was used to maintain anaerobic conditions. The apparatus has been used to grow *Selenomonas ruminantium*. A more detailed description of the apparatus with minor improvements was reported by Hobson (25).

King and Vessal (26) reported on the preparation and specific action of several cellulases derived from *Trichoderma viride, Cellvibrio gilvus* and *Myrothecium verrucaria*.

Leatherwood (27) reviewed cellulase production in *Ruminococcus albus*. Further work on *Eubacterium (Cillobacterium) cellusolvens* was reported by Prins et al (28). Isolation of the type strain and determination of its products was first presented by Hungate (1). Prins et al reported the isolation of seven strains of *Eubacterium cellusolvens* and its products on a cellulose medium. The main fermentation products of five of the strains were hydrogen, carbon dioxide, formate, butyrate and lactate, with small amounts of propionate. The strains were all able to carry the fermentation of cellulose to near completion. However, because a ball-milled cellulose was used, it is difficult to determine how much of the cellulose was crystalline and how much was amorphous.

Leatherwood and Sharma (29) reported the discovery of a novel anaerobic cellulolytic bacterium. The organism is able to survive on cotton linters as a sole carbon source during liquid culture, and is capable of being preserved by freezing. The fermentation pattern of the organism is similar to that of *Butyrivibrio fibrisolvens*.

Kingsley and Hoeniger (30) reported their observations on the genus Selenomonas, a group of anaerobic bacteria capable of hydrolyzing cellulose. A total of eight ruminant and three oral Selenomonads were studied. These organisms all grew at relatively low pH, generally growing best between pH 4.5 to 5.0. They could be grown on cellulose, cellobiose/glucose medium or starch. These organisms produce acetic, butyric, formic, lactic and propionic acids from carbohydrates. It is still tentative to assign these organisms to the genus Selenomonas, and it is still unclear whether these organisms are protozoa or bacteria. However, it is clear that they play a role in the digestion of cellulose in the rumen. Rumen Selenomonads form, in addition to organic acids, hydrogen sulfide. Oral Selenomonads do not form this gas. The species names were *Selenomonas ruminantium* and *Selenomonas sputigena*.

Han and Callihan (31) investigated several different pretreatments, including treatment with sodium hydroxide, ammonia, sulfuric acid, crude cellulase, grinding and high-pressure cooking. Only treatment with 4% sodium hydroxide or 5.2% ammonia increased the actual digestibility of the cellulase. Treatment with cellulase and sulfuric acid increased the solubility of the cellulose, but did not increase the actual percentage of degradation. Grinding and high-pressure cooking did not increase either the rate of digestion of the cellulose or the actual percentage degradation. The test organisms used were Cellulomonas sp. and Alcaligenes sp. The preparations were tested by incubation with these microorganisms for a period of 100 hours.

Much of the work which has been performed on anaerobic cellulolytic bacteria has grown out of a desire to improve ruminant fermentation of forages. To a great extent, the work has been characterized by the use of mixed cultures. Much of this work is qualitative, rather than quantitative. The difficulty of culturing large quantities of these bacteria has, to a certain extent, limited the isolation and purification of cellulolytic enzymes. Where purification work has been performed, the very low activities of the product enzymes cast some doubt upon the purity of the substrate and the exact action of the enzymes upon the substrate.

Protozoa

Protozoa were noted by many early investigators in rumen fluid. Although the importance of these organisms in ruminant nutrition is not yet established, work by various investigators has shown that they do have the ability to degrade cellulose, and, in particular, to attack and ingest small cellulose particles. Because of the difficulties encountered in culturing these organisms, little work has been performed in this area. What work has been performed is, to a great extent, qualitative.

Hungate (32) reported on the successful culture of *Euplodinium neglectum*, a rumen protozoan capable of digesting cellulose. The protozoa were grown in a medium consisting of acid-swollen cellulose and mineral salts. In such a cellulose-mineral medium, the protozoa survived best at pH values ranging from 6.1 to 7.6. It was found, however, that cultures of rumen protozoa were better maintained on media containing rumen fluid or fresh-cut grass. Cultures were successfully maintained, using transfers every two days, across periods of up to 22 months. It was not possible to grow the cultures axenically, although this was attempted repeatedly.

It was demonstrated that an extract of the cultured protozoan was capable of hydrolyzing cellulose to produce reducing sugars. The protozoa extract lost the ability to hydrolyze cellulose when boiled. The cellulase-containing extract was found to be most active at pH values between 4.0 and 6.6, with a maximum activity at 5.0. The type of cellulose attacked by the protozoan, together with the products of hydrolysis and metabolism, are in some doubt.

Baker (3) reported on direct microscopical observations of bovine rumen populations. Baker identified several species of protozoa, including Ophryoscolex, Diplodinium, Epidinium, Entodinium, Isotricha and Dasytricha.

Hungate (22) extended the previously mentioned cellulose experiments with *Euplodinium neglectum.* The name of the protozoan had become, by that time, Diplodinium. The culture system mentioned in the earlier report was extended to include experiments of *Diplodinium maggii, Diplodinium multivesiculatum, Diplodinium denticulatum* and *Entodinium caudatum.* Extracts prepared from Diplodinium cultures were able to hydrolyze cellulose to produce reducing sugar, as measured by test with Benedict's solution. Extracts from *Entodinium caudatum* were not able to hydrolyze cellulose. A rapid synthesis of polysaccharide from cellulose was demonstrated in *Diplodinium maggii.*

Hungate (16) summarized research in rumen protozoology. He reported that successful in vitro culture of several genera of protozoa, including Diplodinium, Endodinium and Epidinium, had been performed for periods ranging up to 22 months. However, culture of holotrich protozoa for extended periods of time had not been successful. Successful in vitro experiments on cellulose digestion using Diplodinium were reported.

Coleman (33) reported experiments in the culture and metabolism of rumen ciliate protozoa. Coleman reported the successful culture of *Entodinium caudatum* on a medium of inorganic salts, hay and rice grains. Coleman also reported the degradation of two-thirds of medium cellulose by *Metadinium medium.* The degradation products were not quantified.

Cellulases

Cellulases obviously are produced by anaerobic bacteria and protozoa. The methods whereby these cellulases are demonstrated, and the means used to measure their activity vary considerably. In this field, many of the measures used for the isolation of bacteria and the demonstration of their ability to hydrolyze cellulose were qualitative, rather than quantitative. This appears to have arisen from the difficulty in growing large cultures of anaerobic bacteria, and from the difficulties in the preparation of cellulases.

A further difficulty is presented by the differences within individuals of a given species or strain in their ability to ferment a single substrate. In part, it may have arisen as a tradition from the first attempts to isolate cellulolytic bacteria. However, the bulk of the work on anaerobic cellulose degradation appears to have been performed using qualitative techniques and small cultures.

The successful isolation and cultivation of a relatively pure culture of a rumen organism capable of hydrolyzing cellulose seems to have begun with Hungate's classic isolation of *Euplodinium neglectum* in 1942. The use of cellulose by this organism was demonstrated by experiments illustrating that the culture failed when cellulose was removed as a constituent of the growth medium. Microscopic examination of the protozoa, which disclosed cellulose particles ingested by the protozoan, also was used to determine that it did digest cellulose. A crude extract of the protozoan culture was used to hydrolyze cellulose, and the presence of reducing sugars in enzyme-cellulose mixtures was demonstrated. However, the products were not identified, and the cellulose used was an acid-swollen cotton. The effect of ball milling and of acid-swelling on the cotton was not known, nor was the ability of the protozoan to digest α-cellulose determined.

Using the methods of culture which he developed with *Euplodinium neglectum,* Hungate (22) investigated the production of cellulases by several other anaerobic protozoa. He determined that cell-free extracts of *Diplodinium maggii* and *Diplodinium multivesiculatum* were capable of producing reducing sugars from cellulose. He observed that three protozoan cultures, *Diplodinium maggii, Diplodinium multivesiculatum* and *Diplodinium denticulatum,* ingested cellulose particles from the medium. However, no attempt was made to determine whether the organisms were capable of degrading α-cellulose, or large cellulose fibers. The cellulose used in these experiments was ball-milled acid-swollen cellulose. Excretion of cellulase into the medium was not determined.

Hungate (2) reported the isolation and fermentation characteristic of an anaerobic cellulose-decomposing bacterium, *Clostridium cellobioparus.* He showed that 73% of the input cellulose-carbon could be recovered from a *C. cellobioparus* fermentation as products. The fermentation products included carbon dioxide, hydrogen gas, acetic acid, formic acid, lactic acid and ethanol. Total dissolution of ball-milled acid-swollen cellulose could be obtained if a culture was left at room temperature for several weeks following cessation of active growth. The product of this long-term hydrolysis was cellobiose. This product was also produced in older cultures. When the optimum pH for several culture aliquots was tested, the optimum pH was found to be 5.5. Demonstration of the cellulase was also made using the "clear zone" around a colony in solid cellulose medium.

This was the first use of the technique of selecting cellulolytic organisms using the "clear zone" method of analysis. This method has been used successfully for 30 years. It appears to be the most used method of selection for organisms capable of degrading cellulose, as well as the most common method for demonstrating the presence of an extracellular cellulase.

This method has been used by several subsequent investigators to demonstrate the presence of cellulases, or as a means of identifying cellulolytic organisms. However, there are some difficulties in interpretation of the results. The first difficulty lies in determining the composition of the cellulose used. Different investigators used different methods for the preparation of the cellulose powder used in the roll tube cultures.

Hungate (34) first used a thin layer of cellulose agar in part of a tube during his study of *Micromonospora propionici,* although he did not adopt the use of roll tubes until considerably later [Hungate (12)]. He determined that the colonies formed in thinner sections of rumen fluid agar were easier to select than were those formed in either thick layers of agar or in shake tubes. However, to combat the problem of large particles of cellulose, which were not easily observed during degradation and were not uniform, Hungate (32) used a technique of swelling cotton in hydrochloric acid and subsequently grinding it for several days in a pebble mill. This technique is currently used to varying degrees by most investigators. Unfortunately, it is difficult to determine what happens to the cellulose tested in this fashion. In some cases, investigators used finely divided Whatman cellulose powder. Some used acid-washed powder, while others used standard pure cellulose powder, either Whatman or Solka-floc.

It is difficult to determine from the investigators' description whether the cellulose tested is pure cellulose or a mixture of cellulose and other substances. The cellulose used is swollen or altered so that it is difficult to determine whether the cellulose is α- or β-cellulose, or a mixture of both. This makes it difficult to determine the probable action of cellulolytic organisms on the cellulose used as a substrate.

The clear zone means of identifying cellulolytic organisms has been used by several investigators. These include Hungate, Leatherwood and Sharma and Prins et al.

In addition to the methods of crude extracts and clear zones mentioned above, one other method has been used for the determination of cellulolytic activity by several investigators. This method, although widely used in work on aerobic fungi, is seldom used in anaerobic microbiology. It is the method of placing a small amount of either a culture filtrate or of the culture itself in a tube filled with carboxymethylcellulose gel medium and observing liquefaction of the medium. This method, used by Goto and Okabe (11), has the disadvantage of requiring a second culturing of the organism cultured in order to permit the isolation of a single clone culture. It is also difficult to measure the liquefaction accurately.

Although the clear zone method is used to identify cellulolytic bacteria, it is the least quantitative method for cellulase assay currently in use. The method of testing extracts of cells, or filtrates of cultures, on a variety of substrates is less qualitative. The method was first used by Hungate in his work on *Euplodinium neglectum*. With proper care, it is possible to make the method somewhat quantitative.

Following Hungate's work with *Euplodinium neglectum,* other investigators used crude microbial extracts or culture filtrates to investigate the cellulolytic activity of microorganisms. Most of the work was performed on aerobic fungi, since these organisms caused a considerable supply problem during World War II, and since they were easily cultured in large quantities on defined mineral salts-cellulose media [Reese and Mandels (35)].

Some cellulose degradation studies were performed on ruminants, because of their economic importance. In some cases, investigators used extracts of rumen preparations to determine the degradability of various cellulosic or lignocellulosic materials. Investigators who tested crude rumen extracts on cellulosic materials include Kitts and Underkofler (9), Festenstein (36) and Halliwell (37). These investigators have used various substrates. Kitts and Underkofler used filter paper, Alphacel (a form of α-cellulose), and two types of carboxymethylcellulose. However, their preparations were unable to degrade a significant amount of the added cellulose.

Festenstein used a butanol extract of sheep rumen fluid microorganisms to degrade sodium carboxymethylcellulose and cellobiose. Halliwell used treated cotton fibers, Whatman cellulose powder, swollen cellulose powder prepared from either Whatman cellulose powder or cotton, and dewaxed acid treated cotton.

Halliwell and Festenstein both obtained fairly high percentages of degradation, ranging up to roughly 90% of the input cotton or cellulose powder substrate, in some cases. These authors used a complicated technique to protect their enzyme extracts from air and heat. It appears that this is necessary in order to produce a high-quality extract. Halliwell suggested that the high degradation percentages and rapid rates are due to the mixed culture used. The products of degradation are given as glucose from carboxymethylcellulose and as soluble reducing compounds. Halliwell stated that there is no known method of producing a rumen microorganism extract which has the same vigor against whole native lignocellulose or cellulose shown by a whole rumen culture.

Culture filtrates and crude enzyme preparations having cellulase activity have been used by some investigators. These preparations often did not have a high activity. Halliwell indicated that this might be due to the presence of impurities in cellulosic compounds tested. He suggested that some of the activity labeled "cellulose degradation" might, in reality, be the degradation of the small quantities of hemicellulose or other materials present in cellulose tested. This might, indeed, be the case with swollen, ball-milled or extensively treated cellulose preparation rations. He particularly cautioned against the results of various investigators who used preparations capable of hydrolyzing only a few percent of the available cellulose.

Unfortunately, most of the investigators whose results are described below fall into Halliwell's category of low-yield preparations, or preparations having less activity than the bulk enzyme mix from which they were derived. It is unclear whether the cellulases produced by anaerobic microorganisms are extremely difficult to purify, or whether there are missing activity factors in the preparations tested, or whether a variety of cellulases possible from different sources are required to completely hydrolyze cellulose.

In work prior to 1957, work with enzyme systems was confined to the crude extracts mentioned above. In 1957, Matthijssen (38) reported the purification of cellulolytic enzymes from *Corynebacterium fimi*. The enzyme system was found to have a temperature optimum at 40°C and an optimum pH between 6 and 7. Cellobiose, rather than glucose, was found to be the final hydrolysis product. Enzyme preparations were made using ammonium sulfate, ethanol, acetone and combinations of these precipitants. β-Glucosidase activity was lost in the ammonium sulfate preparations.

Although cellobiose was the final hydrolysis product of the fermentation, it had an inhibitory effect on the operation of the enzyme system when added to an in vitro reaction. Matthijssen was unable to find any oligosaccharides formed as hydrolysis products. He thought that this indicated an end-wise attack of the cellulase. Where it was used, a β-glucosidase isolated from the organism was capable of hydrolyzing cellobiose to glucose. Only a single cellulase, however, was purified.

Storvick and King (18), who isolated and purified several cellulase components from *Cellvibrio gilvus* culture filtrates, found four cellulases capable of hydrolyzing cellulose. When incubated with cellulose oligosaccharides, no cellulase was capable of hydrolyzing either cellobiose or cellotetrose. Cellobiose and glucose were produced from cellopentose in a 2:1 ratio. Cellobiose only was produced from cellohexose. Each of the cellulases isolated appears to have a pH optimum between 7.0 and 7.5.

It appeared, since cellobiose was the only product of hydrolysis formed from even-numbered oligosaccharides, that these enzymes all cleave cellulose in two-glucose units, starting at a strand end. The only problems which King and Storvick appeared to have with these preparations were the presence of an inhibitor, as shown by the increase in enzyme activity during dilution; the loss of a part of the cellulase complex, as shown by an abrupt increase in activity following addition of boiled rumen fluid; and a very low activity in the preparations used.

It is reassuring that the hydrolysis products of the cellulases tested in both cases were cellobiose. This is in agreement with Hungate's finding (6) that cellobiose is the main rumen sugar, and appears to be the main saccharide produced by rumen cellulolytic bacteria. In pure cultures of cellulolytic rumen bacteria, however, cellobiose must be converted to other products, including lactic, acetic and butyric acids, carbon dioxide and hydrogen. Most rumen cellulolytic bacteria are capable of using cellobiose as a sugar, but many are not capable of using glucose.

King explored *Cellvibrio gilvus* cellulases further (26). They used oligosaccharides as a means of determining the precise action of cellulases isolated from Cellvibrio. They did this because hydrogen-bond energy calculations showed that random attack by a cellulase would produce products consisting primarily of cellobiose, as larger fragments would be retained in the cellulose crystal structure by hydrogen bonding forces. Although small oligosaccharides cannot be expected to simulate the larger polymers, they could give products which varied with respect to enzyme action. King and Vessal determined that the degree of ionization of a substrate was an indication of its resistance to enzymatic attack, or, conversely, that enzyme hydrolysis was proportional to the degree of unionization of the material used, carboxymethylcellulose.

King states that *Myrothecium verrucaria* produces a random cleaving endoglucanase, while *Cellvibrio gilvus* primarily produces an endwise-cleaving cellulase. The hydrolysis of cellulose by *Cellvibrio gilvus* appears to produce primarily cellobiose and cellotriose.

More research in the area of cellulase production and the mechanisms of anaerobic cellulose degradation is needed. Although there has been considerable work in the area of anaerobic utilization of cellulose, most of the work has been highly qualitative. This appears to have been due to the difficulty of growing anaerobic cellulolytic bacteria and to the difficulty of isolating their extracellular enzymes in such a fashion that they retained their activity. Even in current work, such as that of King (26), where considerable attempt has been made to purify cellulase components from anaerobic cellulolytic bacteria, the results have not been very successful.

For example, the cellulases isolated by King and Vessal have very low reaction rates. Little investigation of the required cofactors for anaerobic cellulases, or of the control of their biosynthesis, has been performed. If these systems are going to be used effectively as a means of preparing intermediates, or for fuel synthesis, it is important that their biochemistry be elucidated.

Chemical Products

Few quantitative results were found concerning the specific yield of products from the pure culture degradation of cellulose. This is due, in part, to the recent development of anaerobic microbiology as a science, together with concurrent techniques for the pure culture of anaerobic microorganisms. It is also due to the difficulty in preparing, growing and analyzing large bench-scale degradation tests of various pure anaerobic cultures. There are, however, some available quantitative results on the fermentation of cellulose by pure cultures. There is also a large volume of material on the degradation of various fodders by a mixed rumen culture, since these types of experiments are used to test the degradability of various fodders in connection with feeding experiments. Because the feeds tested are not well-defined chemical compounds, like pure cellulose, the results may not be readily translated to yield per unit cellulose, but are given in a form which permits them to be translated into amounts of volatile acids per unit liquid volume.

Table 2.1 indicates that several different types of volatile acids are formed during the fermentation of fodder. Although the fermentation products vary considerably with time and with the type of substrate, it appears that it is possible to produce acetic, propionic and butyric acids from a variety of substrate materials at a good yield. It is important to remember that, other than the milling which is done for the preparation of a uniform sample, no extensive pretest treatment of the fodder material is performed. Thus, the process considered would have primarily the preprocess expenses of hauling and postprocess expenses of waste disposal, but not have the preprocess expenses of extensive preparation.

Table 2.1 also shows the possible yield of various volatile acids which are routinely produced in the rumen during fermentation of various foodstuffs. The samples tested were incubated in an artificial rumen, a small, anaerobic fermentation vessel closed with a bunsen valve, which was seeded with a mixture of rumen fluid and a buffer containing mineral salts. The mixture was incubated at 39.5°C for a period of 12 hours (39). The data obtained from artificial rumen fermentations may closely approximate the expected industrial results using a batch fermenter and a mixed microbial cellulose degradation culture. It is interesting to note that the culture produced more acetic than any other volatile acids.

TABLE 2.1: VOLATILE ACIDS FORMED BY THE FERMENTATION OF 1 KG OF MIXED HAY/CONCENTRATE FEED*

Material	Equivalents per Kilogram of Organic Material Digested ...Rumen Fluid from...	
	Cow 1	Cow 2
Straight hay plus concentrates		
Acetic acid	3.2	4.5
Propionic acid	1.0	1.3
Butyric acid	0.8	1.1
Higher acids (as valeric)	0.6	0.6
Ground hay plus concentrates		
Acetic acid	3.9	4.4
Propionic acid	2.3	2.5
Butyric acid	1.2	1.1
Higher acids (as valeric)	0.6	0.7

*Adapted from Balch, Reference (39).

Source: ORNL-5056

CONTROLLING FACTORS IN METHANE FERMENTATION

Investigative work on determining the controlling factors in methane fermentation has been done by R.E. Speece and R.S. Engelbrecht at New Mexico State University (PB 216 556).

The ultimate goal of research by Speece and Engelbrecht has been to determine the chemical, physical and biochemical requirements which will enable the methane fermentation to proceed at rates on the order of one magnitude greater than commonly observed in conventional anaerobic digestion.

A considerably detailed study was conducted to assay the methane fermentation stimulation potential of six compounds which had shown promise in a previous study. Next, a series of studies was designed to evaluate the influence of many physical and chemical factors involved in digestion. Following this, the methane fermentation requirements for trace organics, such as vitamins and amino acids, was studied. Combinations of individual amino acids and vitamins as well as mixtures, e.g., casein hydrolyzate (enzymatic), yeast extract and fortified vitamin B complex, were assayed for their stimulation potential. The surface charge on the microorganisms was altered by inorganic and organic coagulating agents while observing the acetate utilization rate.

Additional fractionation of complex stimulatory substances such as sewage sludge digester supernatant and rumen fluid is necessary in order to more definitely identify the combination of nutrient requirements which enable methane fermentation to proceed at high rates. Some fractionation work was done at M.I.T. and additional studies are under way on this project. In work done with methane organisms in the field of Dairy Science, a requirement for the higher branched and normal fatty acids, such as methylbutyric, isobutyric, n-valeric and n-butyric has been found. These were obtained from the acidified ether extract fraction.

Cation and anion exchange, solvent extraction, activated carbon adsorption, distillation, chromatographic separation, acid and base hydrolysis, and other fractionation procedures were used both singularly and in combination to more clearly identify the stimulatory agent or agents. Assay for stimulation potential was conducted with sludge taken directly from a sewage sludge digester as well as an enriched culture taken from a washed out stock digester.

Evaluation of Stimulation Potential of Various Stimulants—Part I

The initial phases of the study were designed to evaluate the effects of a number of compounds which had produced stimulation in a previous study. These compounds were: iron, cobalt, thiamine, proline, glycine and benzimidazole. A series of eight 1-liter digesters was set up from a large washed-out stock digester which was purged of the original sludge by daily washing of the contents and replacement by a feed which consisted of acetate, inorganic nutrients and tap water.

The digesters were fed a combination of the compounds according to the following chart:

Digester No.	Fe	Co	Thiamine	Proline	Glycine	Benzimidazole
1	–	x	x	x	x	x
2	x	–	x	x	x	x
3	x	x	–	x	x	x
4	x	x	x	–	x	x
5	x	x	x	x	–	x
6	x	x	x	x	x	–
7	x	x	x	x	x	x
8	–	–	–	–	–	–

The concentration of each compound was fed so as to maintain a concentration of 20 mg/l in the digester. Digesters 7 and 8 served as controls in which all of the compounds and none of the compounds were fed respectively. The substrate was maintained at approximately 1,500 mg/l as acetic acid and fed in the form of the neutral salt calcium acetate. The digesters were operated on a 30-day detention time (DT). The pH was maintained near neutrality by maintaining the alkalinity at 3,000 mg/l as calcium carbonate by the addition of $NaHCO_3$. Additional inorganic salt concentrations in the digester were maintained as follows: 266 mg/l $(NH_4)_2HPO_4$, 27 mg/l KCl and 66 mg/l $MgSO_4$.

Results of First 80 Days of Operation: This set of digesters was then operated according to the above schedule for 80 days and the results are shown in Figure 2.1. The digester which received all of the assay compounds except iron showed a low rate of acetate utilization, 200 mg/l/day, and progressively decreased to about 100 mg/l/day. The digester which received all assay compounds except cobalt exhibited an acetate utilization rate of 200 mg/l/day initially, but rose to approximately 1,000 mg/l/day. The digester receiving no thiamine was retarded for the first 30 days and utilized only about 200 mg/l/day of acetate, but then the utilization rate rose to approximately 600 mg/l/day.

FIGURE 2.1: COMPARATIVE EFFECT OF ADDITIVES—DAY 1 TO 80

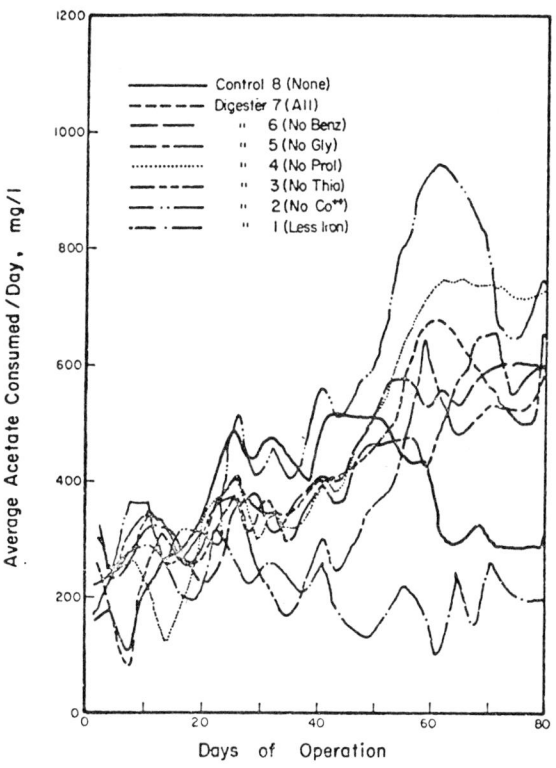

Source: PB 216 556

The digester receiving no proline increased from 200 to a maximum of 700 mg per liter per day. The utilization rate in the digester receiving no glycine increased steadily from 200 to about 700 mg/l/day. The digester receiving no benzimidazole responded similarly to the digester receiving no proline. The control digester receiving all assay compounds responded much the same as the digester receiving no proline. The control digester receiving none of the assay compounds rose from 200 to 500 mg/l/day in the first 45 days and then declined to 300 mg/l/day in the last phase.

The results of this study for 80 days showed a markedly lower acetate utilization for the two digesters receiving no iron and a significantly higher rate for the digester receiving no cobalt. The other digesters operated at about the same rate in a zone between these two extremes.

Results of Operation from Day 81 to 170: Figure 2.2 shows graphically the digestion rates for the period Day 81 to 170. Operation changes were made on Day 81 and Day 90. On Day 81, additions of 5 mg/l of iron were daily added to the digester which received all assay compounds except iron.

FIGURE 2.2: COMPARATIVE EFFECT OF ADDITIVES—DAY 81 TO 170

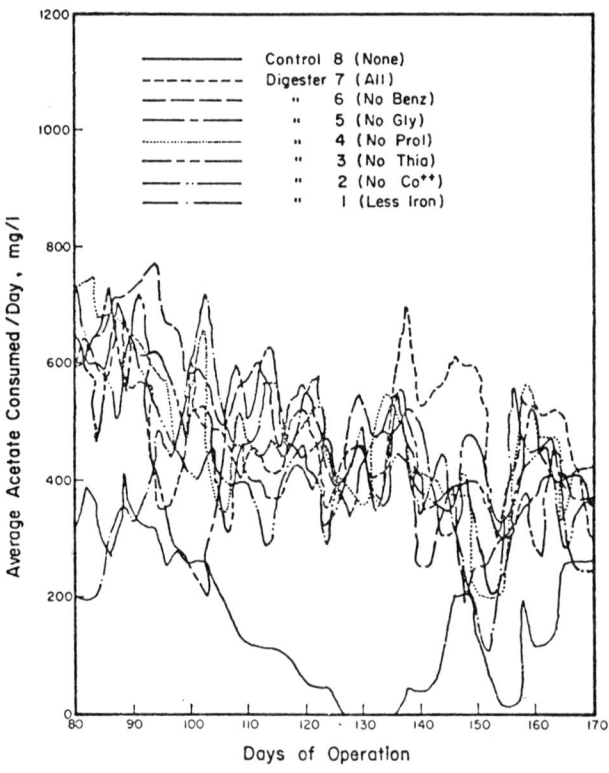

Source: PB 216 556

On Day 90 three changes were made: the concentration of cobalt was reduced from 20 to 1 mg/l in all digesters which received this compound, and the hydraulic detention time in all digesters was reduced from 30 to 10 days. Also, all of the assay compounds were reduced to one-third their original concentration until Day 140 when they were again added at the original concentration. The following table shows the concentration of compounds added to the respective digester feeds, expressed in mg/l:

Digester No.	Fe	Co	Thiamine	Proline	Glycine	Benzimidazole
1	1.67	6.7	6.7	6.7	6.7	6.7
2	1	-	1	1	1	1
3	6.7	6.7	-	6.7	6.7	6.7
4	6.7	6.7	6.7	-	6.7	6.7
5	6.7	6.7	6.7	6.7	-	6.7
6	6.7	6.7	6.7	6.7	6.7	-
7	6.7	6.7	6.7	6.7	6.7	6.7
8	-	-	-	-	-	-

A very marked increase in acetate utilization was evident when iron was added to the digester which previously received all assay compounds except iron. The rate increased from 200 to 720 mg/l/day in the period from Day 80 to 100. The digester receiving none of the assay compounds decreased progressively to nil utilization by Day 130 and was discarded. From Day 140, when the concentration of assay compounds was tripled back to the original level, the digester receiving all compounds operated at a slightly higher rate than the other digesters.

Generally speaking, during this period, all digesters except the one receiving none of the assay compounds operated at approximately the same utilization rate.

Results of Operation from Day 170 to 255: Figure 2.3 illustrates the results during this period. Since the acetate utilization rate in all of these digesters was limited to approximately 400 mg/l/day, an attempt was made to change the conditions so that higher rates could be achieved.

Therefore, the nutrient salts added along with the feed were altered. Ammonium bicarbonate was substituted for sodium bicarbonate as the principal buffer. This increased the ammonium nitrogen content in the digesters from about 60 to 800 mg/l as nitrogen. From Day 200, the concentration of the basic mono- and divalent cations was increased.

There was no noticeable effect in the substitution of ammonium bicarbonate for sodium bicarbonate and the consequent increase in ammonium nitrogen concentration. A minor increase from approximately 300 to 500 mg/l/day of acetate utilization was noted when the mono- and divalent cation concentration was increased.

However, during this entire study of 255 days, one or more limiting conditions existed in all digesters which prevented the achievement of significantly high rates of acetate utilization.

FIGURE 2.3: COMPARATIVE EFFECT OF ADDITIVES—DAY 171 TO 250

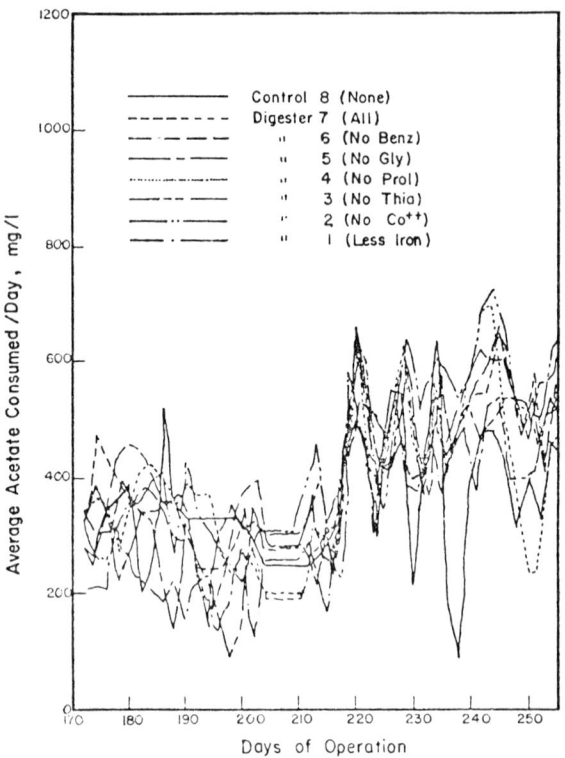

Source: PB 216 556

Evaluation of Stimulation Potential of Supposed Stimulants—Part II

Eighteen digesters were inoculated with 500 ml of seed from a washed-out digester and diluted to 1½ liters with buffered nutrient water. The stimulation potential of the compounds which had produced stimulation in a previous study were assayed individually and in combinations. These digesters were operated for 170 days.

Period 1: From Day 1 to 75, the digesters were operated on a 33-day DT with the nutrient salt concentration in the feed maintained as follows: 140 mg/l $MgSO_4$, 30 mg/l KCl, 300 mg/l $NH_4H_2PO_4$ and 5,000 mg/l $NaHCO_3$. The digesters were fed a combination of compounds at 20 mg/l each during Period 1 according to the schedule at the top of the next page.

Period 2: From Day 75 to 95, the detention time was reduced from 33 to 23 days. The concentration of assay compounds and nutrient salts in the feed was maintained the same as during Period 1.

Bioconversion Mechanisms

Digester	Fe^{+++}	Co^{++}	Thiamine	Proline	Glycine	Benzimidazole	Al^{+++}
A Control	–	–	–	–	–	–	–
B Control	–	–	–	–	–	–	–
C	x	–	–	–	–	–	–
D	–	x	–	–	–	–	–
E	–	–	x	–	–	–	–
F	–	–	–	x	–	–	–
G	–	–	–	–	x	–	–
H	–	–	–	–	–	x	–
I Control	–	–	–	–	–	–	–
J	x	x	–	–	–	–	–
K	–	x	x	–	–	–	–
L	–	x	x	–	–	x	–
M	–	x	x	–	x	x	–
N	–	x	x	x	–	x	–
O Control	–	–	–	–	–	–	–
P	–	x	x	–	x	–	–
Q	–	x	x	x	–	–	–
R	–	–	–	–	–	–	x

Period 3: From Day 95 to 125, the detention time was reduced from 23 to 10 days. The concentration of assay compounds and nutrient salts in the feed was maintained the same as during Period 1.

Period 4: From Day 125 to 170 the iron concentration in digesters C and J was reduced from 20 to 5 mg/l. Also 5 mg/l of iron was added to the feed of the remaining digesters with the exception of the four control digesters A, B, I and O. The concentration of cobalt was reduced from 20 to 1 mg/l in the digesters which were scheduled to receive cobalt in the feed. The detention time remained at 10 days.

The following tabulation shows the rate of acetate utilization during the respective periods:

Acetate Utilization Rate (mg/l/day)

Digester	Period 1	Period 2	Period 3	Period 4
A	206	261	248	198
B	230	284	248	179
C	309	337	284	183
D	117	56	103	121
E	173	213	222	145
F	193	273	231	164
G	233	249	211	169
H	187	143	200	185
I	186	229	233	211
J	198	340	383	358
K	146	210	118	104
L	83	77	118	93
M	47	17	103	97
N	69	32	115	102
O	199	197	268	225
P	80	100	101	94
Q	165	150	95	190
R	234	254	237	185

These results are plotted in Figure 2.4 on a bar graph showing the respective assay compounds in each digester.

40 Energy from Bioconversion of Waste Materials

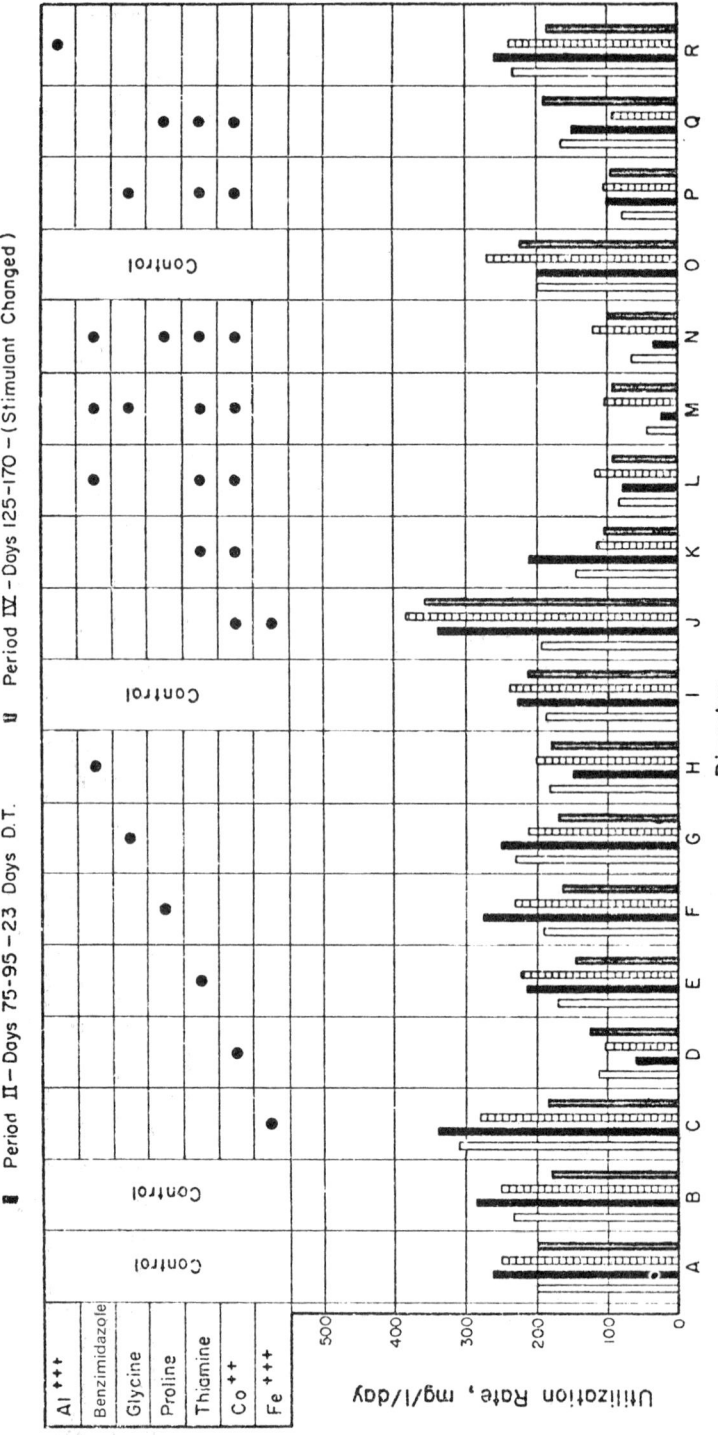

FIGURE 2.4: BAR GRAPH COMPARISON OF ADDITIVES

Source: PB 216 556

Discussion: The significant effects during the operation of each digester are noted as follows:

Digester C operated significantly higher than the control during the Period 1. However, during Period 2 and 3, it operated only slightly higher than the control and operated comparably to the control during Period 4.

Digester D operated consistently below the control. The change during Period 4 increased the rate to about comparable with the control. Digester E decreased to a rate below the control during Period 4.

Digesters F, G and H operated at or slightly below the control during all four periods. Addition of iron in Period 4 had negligible effect.

Digester J operated comparably to the control during Period 1 and significantly higher during Period 2. A marked increase in rate above the control occurred during Period 3 when the detention time was cut to 10 days. However, when the iron was reduced from 20 to 5 mg/l and the cobalt was reduced from 20 to 1 mg/l, the rate fell to that of the control.

Digester K operated generally below the control. Very little change was noted during Period 4. Digester L operated significantly below the control with very little change during Period 4.

Digester M operated at almost a nil rate during Periods 1 and 2, and then increased to about 50% of the control rate during Period 4. There was no apparent effect during Period 4.

Digester N operated significantly below the control during Periods 1 and 2, but began to approach the control rate during Period 3. There was no apparent effect during Period 4. Digester P operated significantly below the control during all four periods.

Digester Q operated slightly below the control during Period 1, while the rate dropped even lower during Period 2. Period 3 resulted in an almost nil rate, but by adding the 5 mg/l of iron and reducing the cobalt from 20 to 1 mg/l in Period 4, there was a steady increase in rate until it exceeded the control. Digesters Q and R operated comparably to the control during all periods.

The stimulatory effect of iron is apparent as noted by the acetate utilization rates in Digester C during Periods 1, 2 and 3 when it was added at a concentration of 20 mg/l in the feed. The other obvious stimulation occurred in Digester J during Periods 2, 3 and 4. This digester received a combination of iron and cobalt.

The inhibitory effect of 20 mg/l of cobalt in the feed is marked during Periods 1, 2 and 3 in all the digesters which received cobalt except Digester J which also received iron. Many of these digesters operated at less than 50% of that of the control.

The addition of thiamine, proline, glycine and benzimidazole slightly retarded the rate of acetate utilization when added singularly to a digester. It appears the addition of thiamine overcame the toxic effect of cobalt when added in this combination in Digester K as compared to Digester D during Period 2. Proline was slightly more stimulatory than glycine.

Effect of Inert Surface on Acetate Utilization Rate

Two replicate digesters were set up to evaluate the stimulatory effect on acetate utilization of finely ground asbestos. One digester received 30 grams per liter of finely ground asbestos on the first day only while the second served as a control and received none. The seed used to inoculate both digesters came from a washed out stock digester being fed calcium acetate. The feed to both digesters contained the same concentrations of nutrient ions Na, NH_4, K, PO_4, Mg and sufficient acetate to maintain between 500 and 1,500 mg/l of volatile acids in the digester at all times. Both digesters were operated on a 10-day detention time and mixed once per day. The duration of this study was 40 days.

Results: For the first 10 days, both digesters operated at the same rate. Thereafter until the end of the study, the rate of utilization in the digester which received asbestos operated at a rate approximately double that of the control digester as shown in the accompanying graph. The digestion rate in the control digester fell from about 700 to 350 mg/l/day by the end of the run, while the digester with asbestos operated at a rate between 900 and 1,000 mg/l/day during the period. These results are shown in Figure 2.5.

FIGURE 2.5: EFFECT OF ASBESTOS ON DIGESTION RATE

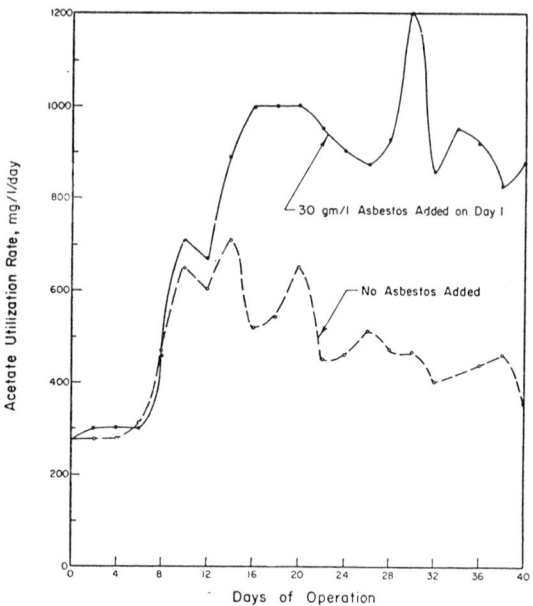

Source: PB 216 556

It should be noted that it was not possible to withdraw a uniform sample of the contents from the digester containing asbestos. This may have resulted in less wastage of cell mass from this unit as compared to the control and accounted for part of the increased digestion rate.

Effect of Asbestos, Asbestos Extract, Iron and No Additive on Acetate Utilization Rate

Since the presence of asbestos in a digester resulted in increased acetate utilization, an attempt was made to evaluate more closely the phenomena involved. Asbestos contains certain minerals such as calcium, magnesium, iron and manganese. It was hypothesized that the stimulatory effect of asbestos resulted from the extraction of required minerals from the asbestos under anaerobic conditions. Therefore, four replicate digesters were set up from a stock washed-out digester. One digester received only substrate and nutrient salts and tap water with no additives in the feed solution and thus served as a control. The second digester received 20 g/l of asbestos on the first day. The third digester received the anaerobic extract from asbestos, instead of tap water in the daily feed solution while the fourth digester received 20 mg/l of iron in the daily feed.

These digesters were operated for 35 days and the results are shown in Figure 2.6. It is evident that the digester containing asbestos yielded the highest rate of acetate utilization.

FIGURE 2.6: COMPARATIVE EFFECT OF ASBESTOS, ASBESTOS EXTRACT AND IRON ON DIGESTION RATE

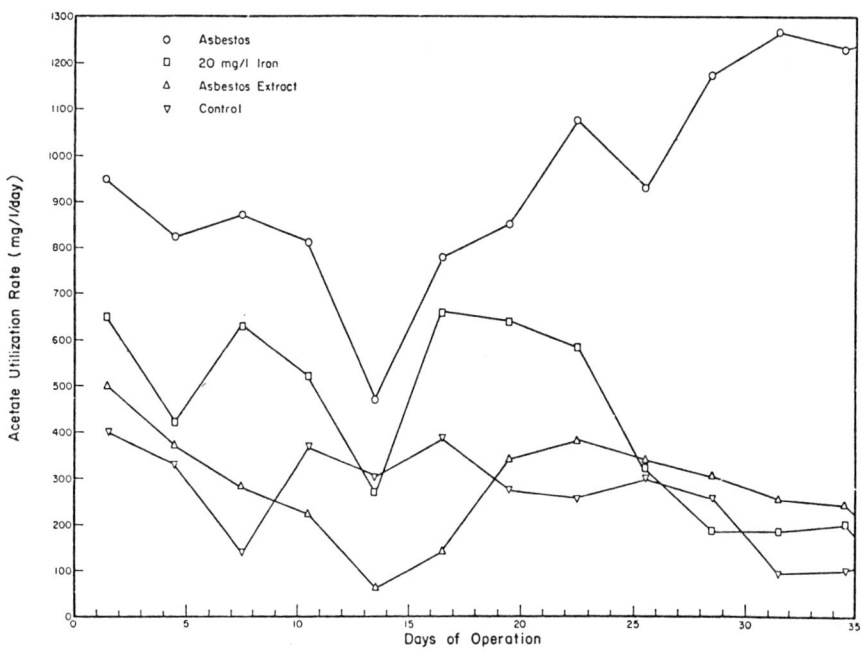

Source: PB 216 556

This would indicate that the function of asbestos is not just a reservoir of iron or other essential minerals which can be leached out under anaerobic conditions. Something in addition, surface area being one factor, is provided by the presence of asbestos.

Effect of CaCO$_3$ Solids on Acetate Utilization Rate

References are found in the literature to bacteria which multiply more rapidly in the presence of a material which increases the surface area, such as beads or an inert, finely divided precipitate. The response of the methane-forming organisms was evaluated in the presence of 0, 1, 5, 10 and 20 grams/liter of CaCO$_3$.

Five replicate 750-ml digesters were set up from a washed-out stock digester. The above concentrations of CaCO$_3$ were added initially and sufficient amounts were present in the daily feed to replenish the inert solids withdrawn daily. Stock nutrients were continually supplied and the acetic acid substrate was maintained at 1,000 to 1,500 mg/l.

Results: Figure 2.7 indicates that the digesters containing the higher concentrations of CaCO$_3$ operated at a higher rate of acetate utilization, especially during the period of Days 15 to 35. Some adverse effect occurred in the digester containing 5 g/l which drastically decreased the rate between Days 6 and 14. However, it recovered from this by the end of the study. Also, air leakage on Day 27 in the digester containing 20 g/l temporarily reduced the rate in that digester.

FIGURE 2.7: EFFECT OF SOLIDS ON DIGESTION RATE

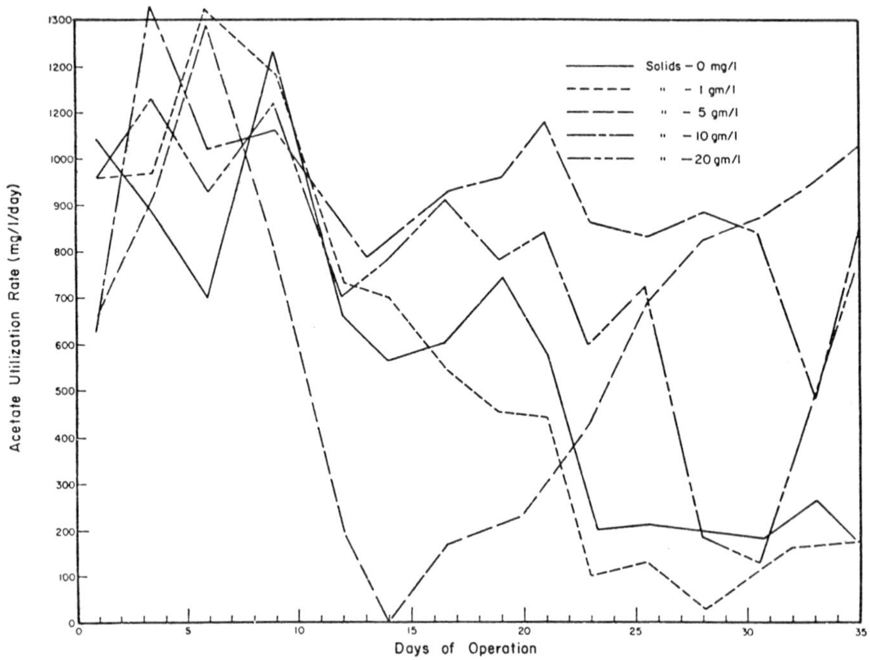

Source: PB 216 556

The mechanism whereby the presence of $CaCO_3$ solids resulted in higher acetate utilization rates is not apparent.

Effect of Volatile Acids Concentration on Acetate Utilization Rate

Two replicate digesters were set up to observe the effect of acetate concentration on utilization rate. Both digesters were initially purged of oxygen with a gas mixture of 50% carbon dioxide and 50% nitrogen, and were then charged with 750 ml of seed from a washed-out stock digester which was fed calcium acetate as the substrate and nutrient salts containing ammonia, phosphate, potassium and magnesium. Iron was initially added to both digesters at the rate of 50 mg/l. No buffer was fed, since the neutral calcium salt of acetic acid was fed as the substrate.

Both digesters were operated identically for three weeks to establish a common baseline of operation. After it was established that they were operating at comparable utilization rates, the volatile acids (VA) concentration in one digester was maintained at approximately 1,300 mg/l, while the volatile acids concentration in the second digester was maintained at approximately 3,500 mg/l. Both digesters were then operated for 60 days.

Results: The operational results are shown in the following table.

Period of Operation (days)	Digester 1 Avg. VA Conc. (mg/l)	Digester 1 Avg. Utilization Rate (mg/l/day)	Digester 2 Avg. VA Conc. (mg/l)	Digester 2 Avg. Utilization Rate (mg/l/day)
1-30	1,360	320	3,780	390
31-60	1,230	190	3,250	540
1-60	1,290	250	3,530	460

It is noted from the above table that during the first month of operation Digester 2 operated at about 120% of the utilization rate of Digester 1. However, during the second month, the utilization rate in Digester 2 was 280% of that in Digester 1. These rates are based on gas production values and therefore are corrected for the higher amount of volatile acids contained in the daily samples withdrawn from Digester 2 which contained the higher concentration of volatile acids.

It appears from this that the microorganisms are dependent on substrate concentration and that the reaction is not of zero order. The higher concentration of volatile acids appears to be responsible for the increase in utilization rate from 390 to 540 mg/l/day in Digester 2 during the second month as compared to a simultaneous decrease in rate from 320 to 190 mg/l/day in Digester 1.

Two factors remain unevaluated in this study: first, the effect of additional precipitated calcium bicarbonate and carbonate in Digester 2 due to the higher feed rate of calcium bicarbonate, and second, the effect of adding 50 mg/l of iron initially, but not replenishing it in the daily feed. The latter was an oversight in this study. It should have been added daily.

Effect of Detention Time on Acetate Utilization Rate

Two replicate digesters were set up from a washed-out stock digester. One digester was operated on a 6-day detention time while the second was operated on a 100-day detention time. The same concentration of nutrient salts was maintained in the feed for both digesters. These digesters were operated over a 45-day period.

Results: The acetate utilization rate at 6 days DT averaged 208 mg/l/day, while at 100 days DT, it averaged 555 mg/l/day. Thus, the overall utilization rate was 2.7 times faster at the longer detention time. However, the organic nitrogen concentration, which is taken as a measure of the bacterial cell mass, averaged approximately 7.0 mg/l in the 6-day DT digester and 60 mg/l in the 100-day DT digester. Using a factor of 9.4 to convert from organic nitrogen to cell mass, this yields 66 mg/l of cells in the 6-day DT digester and 560 mg/l of cells in the 100-day DT digester.

Thus, the acetate utilization rate per gram of cells is as follows:

$$\frac{208 \text{ mg/l/day}}{66 \text{ mg/l of cells}} = 3.15 \frac{\text{g acetate/day}}{\text{g cells}} \text{ @ 6 day DT}$$

$$\frac{555 \text{ mg/l/day}}{560 \text{ mg/l of cells}} = 1.0 \frac{\text{g acetate/day}}{\text{g cells}} \text{ @ 100 day DT}$$

This indicates that a unit mass of cells in the 6-day DT digester utilized approximately 3 times as much substrate as a unit mass of cells in the 100-day DT digester.

The net synthesis of substrate into cell mass was as follows:

6-day DT digester:

$$66 \text{ mg/l cells} \times \frac{1}{6} \text{ wasting} = 11 \text{ mg/l cells synthesized/day}$$

$$\frac{11 \text{ mg cells synthesized/day}}{208 \text{ mg acetate utilized/day}} = 5.3\% \text{ synthesis of acetate into cell material}$$

100-day DT digester:

$$560 \text{ mg/l cells} \times \frac{1}{100} \text{ wasting} = 5.6 \text{ mg/l cells wasted/day}$$

$$\frac{5.6 \text{ mg cells wasted/day}}{555 \text{ mg acetate utilized/day}} = 1.0\% \text{ synthesis of acetate into cell material}$$

Effect of Mixing on Acetate Utilization Rate

A study was made of the effect of mixing on the rate of acetate utilization. Two digesters were set up and operated identically, with one digester being mixed continuously and the other being mixed only once per day at the time of feeding. Both digesters were operated on a 10-day hydraulic detention time and received the usual inorganic nutrients in the feed. Volatile acid concentration was maintained so that a surplus was available except for a number of isolated instances in which the level dropped to less than 500 mg/l. The digesters were simultaneously operated for 90 days.

Results: The digester which was continuously mixed utilized acetate at approximately 2.8 times the rate of the nonmixed digester. The volatile acid utilization rates are shown in the following table for comparison.

	Acetate Utilization Rates (mg/l/day)		
	Average	Maximum	Minimum
Continuously mixed digester	486	720	190
Nonmixed digester	173	450	20

Figure 2.8 is a graphical comparison of the digester performances.

FIGURE 2.8: EFFECT OF MIXING ON DIGESTION RATE

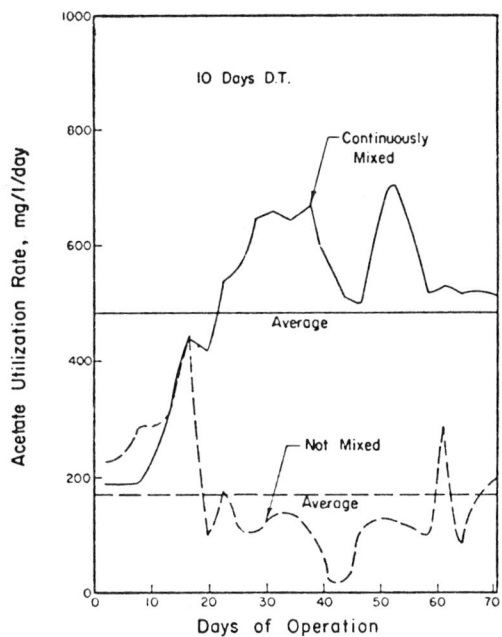

Source: PB 216 556

Assay of Trace Organics for Stimulation of Acetate Utilization Rate

A series of digesters was set up with 100% well-digested sludge from the Champaign-Urbana sewage treatment plant. Trace amounts of various organic compounds, reported to promote the growth of various anaerobic microorganisms, were added on the first day to the digesters in the following combinations:

Digester 1:

Calcium pantothenate	– trace	Nicotinamide	– trace
Adenine	– trace	L-Cystine	– trace
Pyruvate	– trace	β-Alanine	– trace
Thioctic acid	– trace	Benzoate	– trace
Thiamine	– trace		

Digester 2:

Calcium pantothenate	– 10 mg/l
Pyruvate	– 10 mg/l
Adenine	– 10 mg/l

Digester 3:

Menadione	– 10 mg/l
Calcium pantothenate	– 10 mg/l
Vitamin B_{12}	– trace
Biotin	– trace

Digester 4:

Adenine	– 10 mg/l
Thiamine	– 10 mg/l
Calcium pantothenate	– 10 mg/l
Nicotinamide	– 10 mg/l
Thioctic acid	– trace

Sufficient acetic acid was fed daily to maintain the volatile acids concentration above 500 mg/l so that excess substrate was present at all times. Inorganic nutrients were also included in the daily feed. The addition of a fortified vitamin B complex failed to produce stimulation also.

None of these four combinations of trace organic materials was able to produce a noticeable stimulation in the rate of acetate utilization when added to digesters containing 100% well-digested sludge. Nor did the addition of a fortified vitamin B complex produce stimulation either. This indicates that in a sewage sludge digester, these trace organic compounds are not the rate-limiting factor in acetate utilization.

Effects of Surface Charge, Complexing Agent and Other Factors

Surface Charge—$AlCl_3$: A series of digesters was set up to observe the effect of surface charge of the bacteria on the rate of acetate utilization. Five digesters were anaerobically charged from a stock digester which had been purged of the original seed sludge. Acetate and nutrient salts were fed daily to provide substrate and replace the nutrients withdrawn in the daily sample. Aluminum chloride was fed to the respective digesters in the following concentrations: 0, 50, 100, 200 and 400 mg/l.

Results — There was little difference in the acetate utilization of the five digesters. The digester which received 200 mg/l $AlCl_3$ operated during two periods at a rate above the control, but generally operated comparably to the control. The digesters receiving 50, 100 and 400 mg/l $AlCl_3$ operated equally to or less than the control. Figure 2.9 illustrates the comparative rates.

This preliminary study indicates that the effect of making the bacterial surface charge more positive does not appreciably increase the rate at which the anionic acetate substrate is utilized. It does indicate that the addition of a coagulant for removal and recycle of bacterial mass in a waste stream is feasible, since the solids settled very rapidly, leaving a clear supernatant without appreciably diminishing the rate of digestion.

FIGURE 2.9: EFFECT OF COAGULATION ON DIGESTION RATE

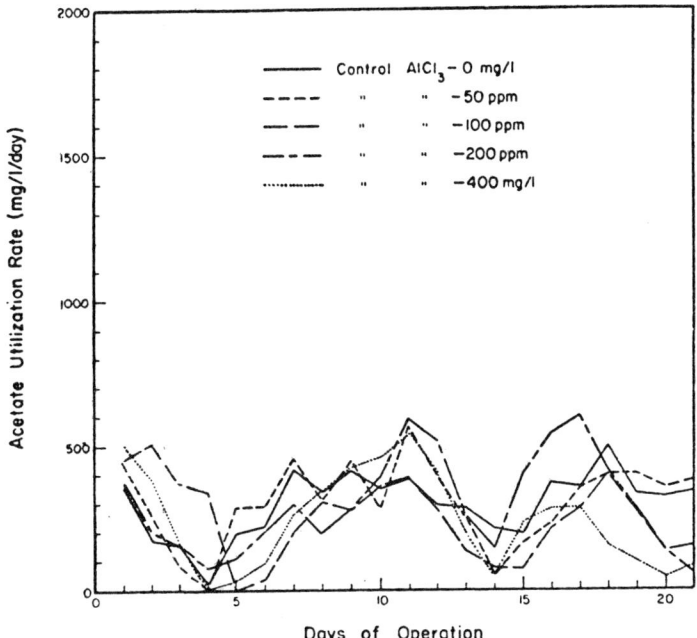

Source: PB 216 556

Surface Charge—Cationic Polyelectrolyte: A cationic polyelectrolyte was added to a digester which had been purged of essentially all of the original seed sludge. Traces of this coagulant were added on Days 1, 7 and 14. From Days 20 to 33, it was added along with the daily feed.

Results — Figure 2.10 illustrates there was no significant stimulation or repression of acetate stimulation due to the polyelectrolyte, except after about 10 days of daily addition when the rate gradually started to decrease. The digester had been continually agitated on a shaking table, but as soon as the shaking was stopped, the solids immediately dropped to the bottom, forming a dense sludge layer. The supernatant appeared comparable to tap water by visual comparison. Again, this preliminary study indicated that bacterial surface charge had little effect on acetate utilization.

Complexing Agent—EDTA: The addition of EDTA to a digester, purged of the original seed sludge, was observed and the result is shown in Figure 2.11. A slug addition of 100 mg/l was made on Day 14 and daily additions were started on Day 20. Figure 2.11 shows the acetate utilization rate vs time. It is difficult to draw any conclusions from this preliminary study. The effect of EDTA on the acetate utilization rate is questionable. There was no significant effect after the slug addition of EDTA on Day 14 and an initial decrease followed by a subsequent increase in rate after daily additions of EDTA were commenced.

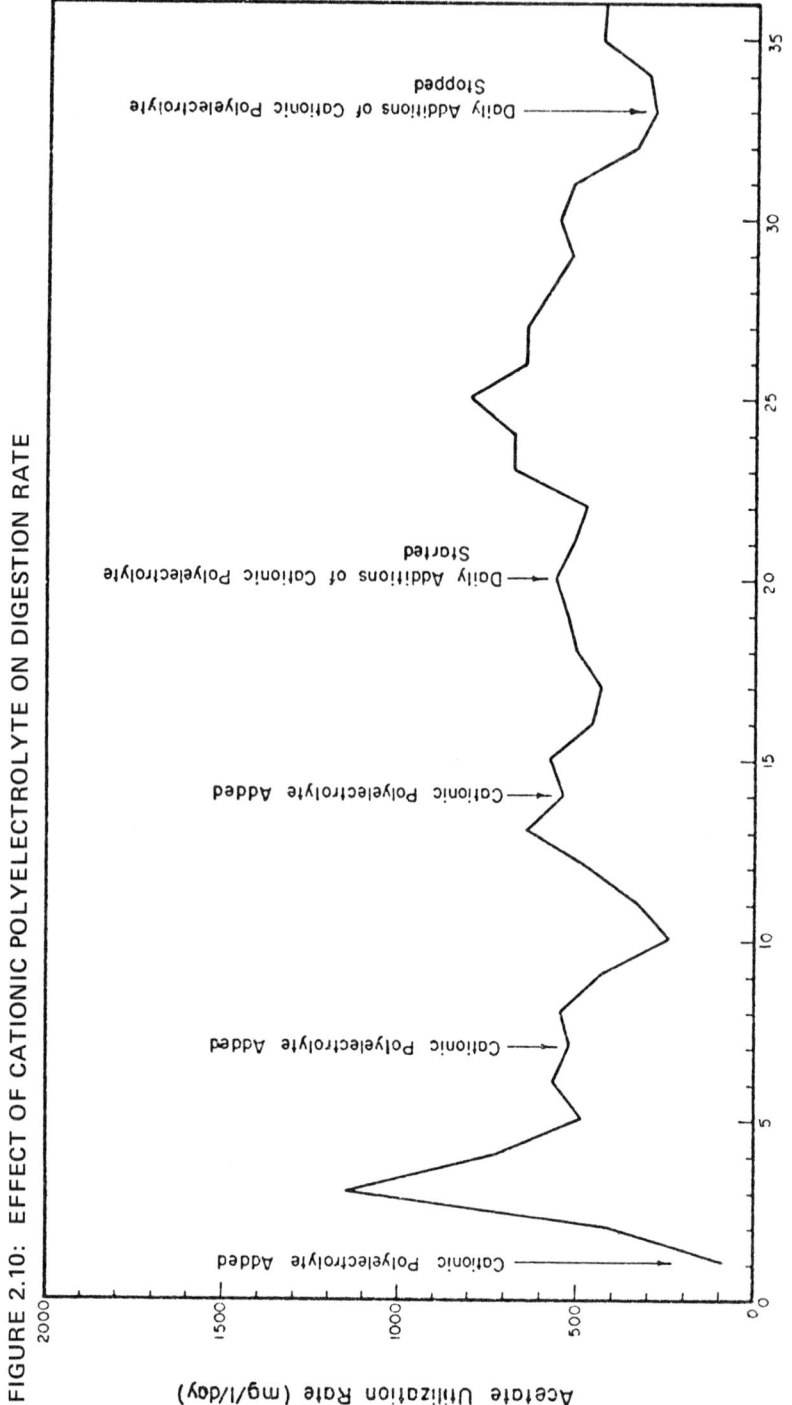

FIGURE 2.10: EFFECT OF CATIONIC POLYELECTROLYTE ON DIGESTION RATE

Source: PB 216 556

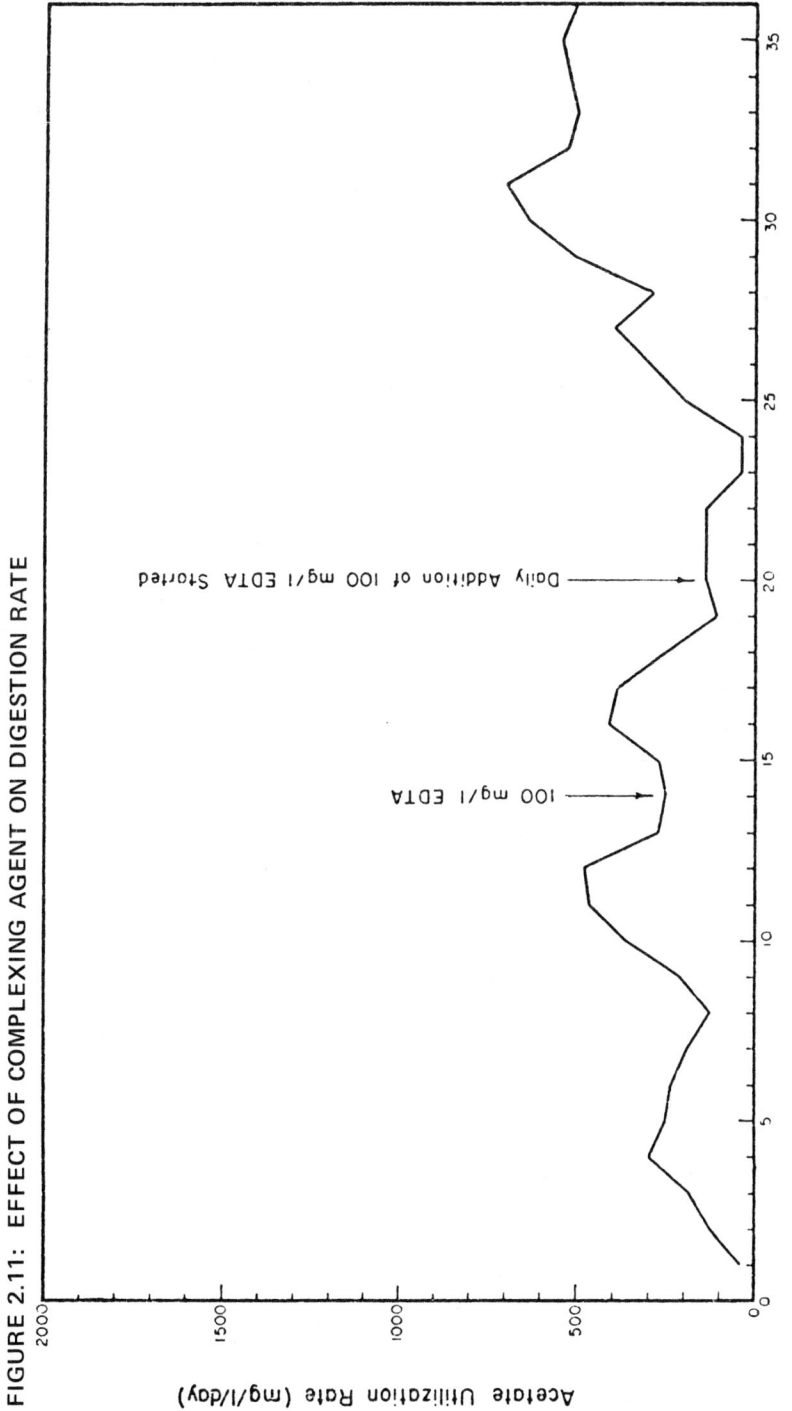

FIGURE 2.11: EFFECT OF COMPLEXING AGENT ON DIGESTION RATE

Source: PB 216 556

It should be pointed out that this study indicated primarily the effect of EDTA itself on the microorganisms. There would be little opportunity for the complexing action to become evident, since any materials which had precipitated from solution would not be returned to solution after the EDTA addition, and thus complexed. The EDTA should be added before any critical materials have been removed from solution.

Effect of Alfalfa Hay Ash: Since methane organisms are found in the rumen of cattle, the beneficial effect of minerals found in alfalfa hay was assayed. 50 g of hay were ashed and added to a digester on two occasions. No significant stimulation in acetate utilization occurred.

Effect of Egg Albumin: The effect of a complex, water-soluble protein was determined by adding egg albumin to a digester. The purpose of this was to pick a protein with a large number of charged sites which were thus capable of altering the charge on the microorganisms. There was no significant effect on the rate of gas production.

Effect of Cyanide: Vitamin B_{12} has been found to be present in anaerobic digestion by a number of workers. Since a cyanide ion is incorporated in the molecular structure of Vitamin B_{12}, the effect of adding small amounts of cyanide to a digester was evaluated. One very insoluble form of cyanide was added, $K_3Fe(CN)_6$, and another more soluble form, KCN, was added in another instance. Toxicity resulted from the addition of 10 mg/l KCN while no significant changes in acetate utilization were apparent after adding 10 mg/l $K_3Fe(CN)_6$.

Effect of Frequent Temperature Variation on Methane Production

The methane-forming microorganisms are generally considered to be more sensitive to physical and chemical changes than the acid-forming microorganisms involved in the anaerobic digestion process. Also, because the methane-forming microorganisms utilize the volatile acid end products produced by the acid-forming microorganisms, failure of the methane-formers to utilize the volatile acids at approximately the same rate as they are produced, can result in a stuck digester.

As a result, the fastidious nature of the methane-forming organisms combined with the critical position they occupy in the anaerobic digestion scheme makes it very important that satisfactory environmental conditions be maintained for them in order to promote good digestion. However, in normal anaerobic digestion, the volatile acid concentration is low, indicating that the methane-forming organisms are capable of utilizing the volatile acids at least as fast as they are being formed. In other words, the acid formation is the rate-limiting step in normal digestion.

A number of studies have been made of the effect of temperature on the anaerobic digestion process (40)(41)(42). These studies have been made using a complex feed such as primary sludges. When using complex substrates, both the acid-formation and methane-formation rates are involved and the net overall effect being observed is actually controlled by whatever the rate-limiting step in the process happens to be. If the volatile acid concentration is low, then the acid-forming step determines the overall rate which is observed.

A study was carried out to determine the effect on the methane-forming organisms in a mesophilic sludge under the following conditions:

1. A temperature change sustained for a number of hours as would occur if the temperature of the digester contents rose or fell over a 24-hour interval.
2. A temperature drop for a 15-minute interval as would occur when digested sludge is premixed with the raw sludge before it is pumped to the digester.
3. A temperature drop for a 2-hour interval as would occur if a proposed anaerobic contact stabilization process were feasible.

Cold, diluted wastes are uneconomical to treat by anaerobic digestion due to excessive heat requirements. However, if the wastes could be adsorbed on a bacterial or inert surface and thus concentrated, it may prove economical to heat this more concentrated form of the sludge and stabilize it by anaerobic digestion. Such a system is shown schematically in Figure 2.12.

FIGURE 2.12: PROPOSED ANAEROBIC CONTACT STABILIZATION PROCESS

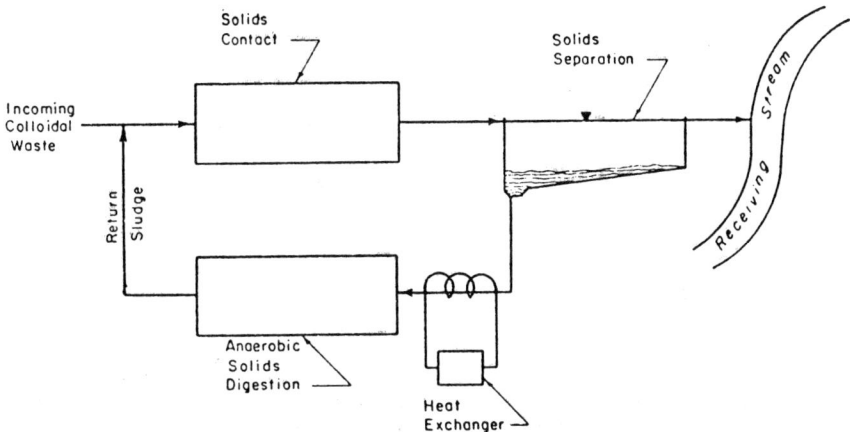

Source: PB 216 556

Procedure: A 2-liter Erlenmeyer flask was placed in a temperature-controlled bath as shown in Figure 2.13. The digester was purged with nitrogen gas and 1.8 liters of sludge from a well-operating 35°C digester was transferred to the flask. The digester contents were mixed by a magnetic stirrer placed underneath the bath. The temperature of the bath was controlled by a thermoregulator and it was kept in circulation by an air diffuser.

At the beginning of a run, the volatile acid concentration was determined. Sufficient acetate was then added to bring the volatile acid concentration to 700 to 1,500 mg/l. Acetate was continuously fed to the digester with an electrolytic pump at approximately the same rate at which it was being utilized. Thus, feed concentration never limited the rate of gas production.

FIGURE 2.13: SCHEMATIC DIGESTION APPARATUS

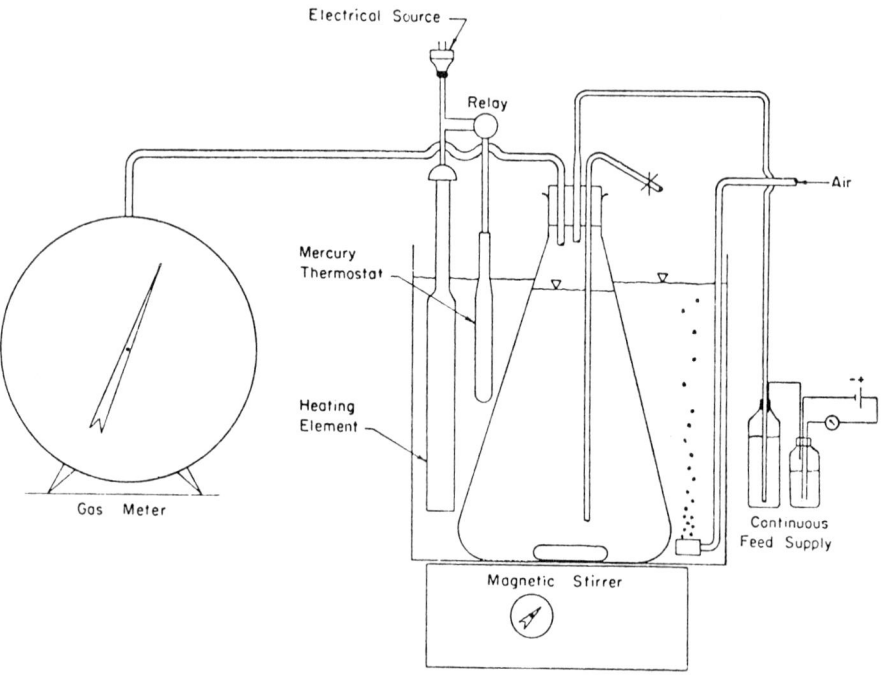

Source: PB 216 556

Since acetate is the predominate volatile acid present in a digester, this was used as the sole energy source. Since only the methane-formers can use acetate under anaerobic conditions, only the rate of activity of these organisms was observed.

At the start of a temperature drop study, the temperature was maintained for about an hour at the level to which it was to be raised after the drop. Gas production was recorded at 10-minute intervals during this first hour, during the drop, and for a sufficient period after the drop to reach equilibrium. These readings were continued during the temperature drop.

Test runs were made to observe the effects of both temperature increases and decreases on the rate of gas production. Observations were made on gas production for the temperature schedules as shown on the following page.

Initial Temperature (C°)	Drop Temperature (C°)	Duration of Temperature Drop (minutes)
35	10	15
35	10	120
35	20	15
35	20	120
50	10	120

Feeding calcium acetate as the sole substrate results in a gas composition of 75% methane and 25% carbon dioxide. The CO_2 introduces an error into the gas production values with each temperature change, however, due to its high solubility in water and the great dependence of CO_2 solubility on temperature. As a result, following each temperature increase, there is an initial marked increase in gas production, after which gas production comes to an equilibrium at a somewhat lower rate. This spike in the gas production is accounted for by the decreased solubility of CO_2 at higher temperatures and therefore the release of CO_2 from solution.

A correction was not made because the exact percentage of CO_2 in the digester atmosphere would have to be known both during the temperature drop when CO_2 would be somewhat above the theoretical 25%. The CO_2 equilibrium of the digester contents and atmosphere was established soon after a temperature increase occurred because the volume at the top of the flask was small and it was rapidly flushed out by the gas evolved.

Figure 2.14 shows the clear response of gas production from mesophilic sludge which had been incubated at 35°C and was incubated for a number of hours at the temperature shown.

FIGURE 2.14: RELATIVE GAS PRODUCTION vs TEMPERATURE

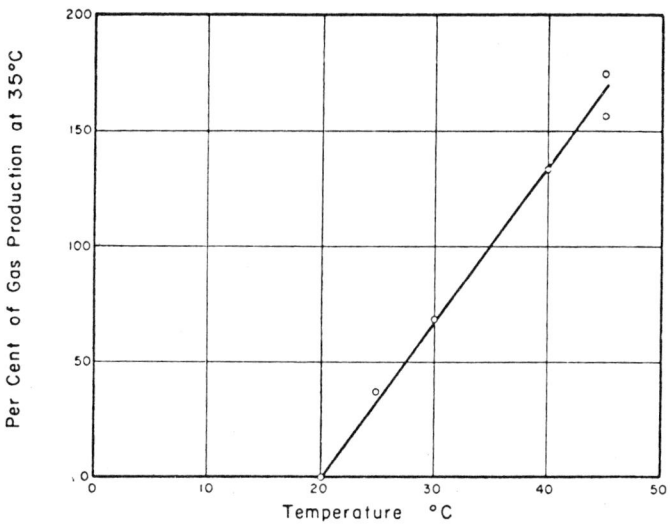

Source: PB 216 556

This figure indicates gas production bears a linear response to temperature as opposed to the traditional logarithmic response. Below a temperature threshold of 20°C, 68°F, gas production was insignificant for the periods observed. This observation is also borne out by later figures.

From this figure, it is seen that a decrease in digester temperature from 37°C to 30°C reduces the gas production to about 50%. Golueke (40) found that the destruction of volatile matter decreased from 50 to 40% when the digestion temperature was decreased from the range of 35° to 55°C down to 30°C. Thus, while volatile matter destruction at 30°C decreased to only 80% of that at 37°C, gas production would suffer much more severely by decreasing to 50% of the original.

As shown in Figure 2.15, the response of the methane-formers to a temperature change is immediate. As the temperature was dropped over an hour's period of time from 34° to 27°C, the gas production decreased to about 12% of that at 34°C. As the temperature was increased from 27°C back to 33°C, the gas production resumed at approximately 100% of that before the temperature drop occurred. A subsequent increase in temperature to 40°C was accompanied by gas production rates of about 140% of those at 34°C.

FIGURE 2.15: DEPENDENCE OF GAS PRODUCTION ON TEMPERATURE

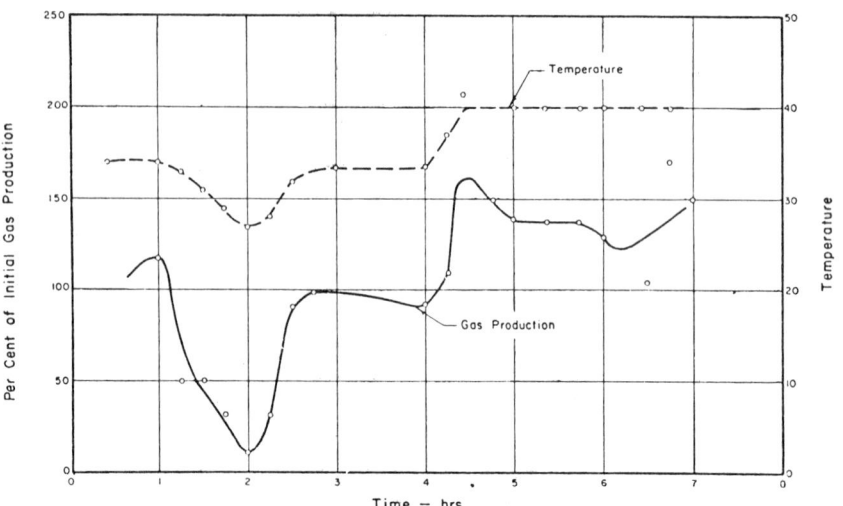

Source: PB 216 556

Figures 2.16 through 2.19 indicate that as long as the digester temperature is below 20°C, gas production is insignificant for the periods shown. However, as soon as the temperature is raised back to the normal range for anaerobic digestion, gas production resumes at a rate proportional to the temperature within the range 20° to 45°C. There appears to be no carryover of the adverse effect on the methane-formers from exposure to temperatures of 10° and 20°C for 15 minutes. Gas production resumed at essentially the same rate as the temperature was restored to the initial level.

FIGURE 2.16: GAS PRODUCTION vs TEMPERATURE

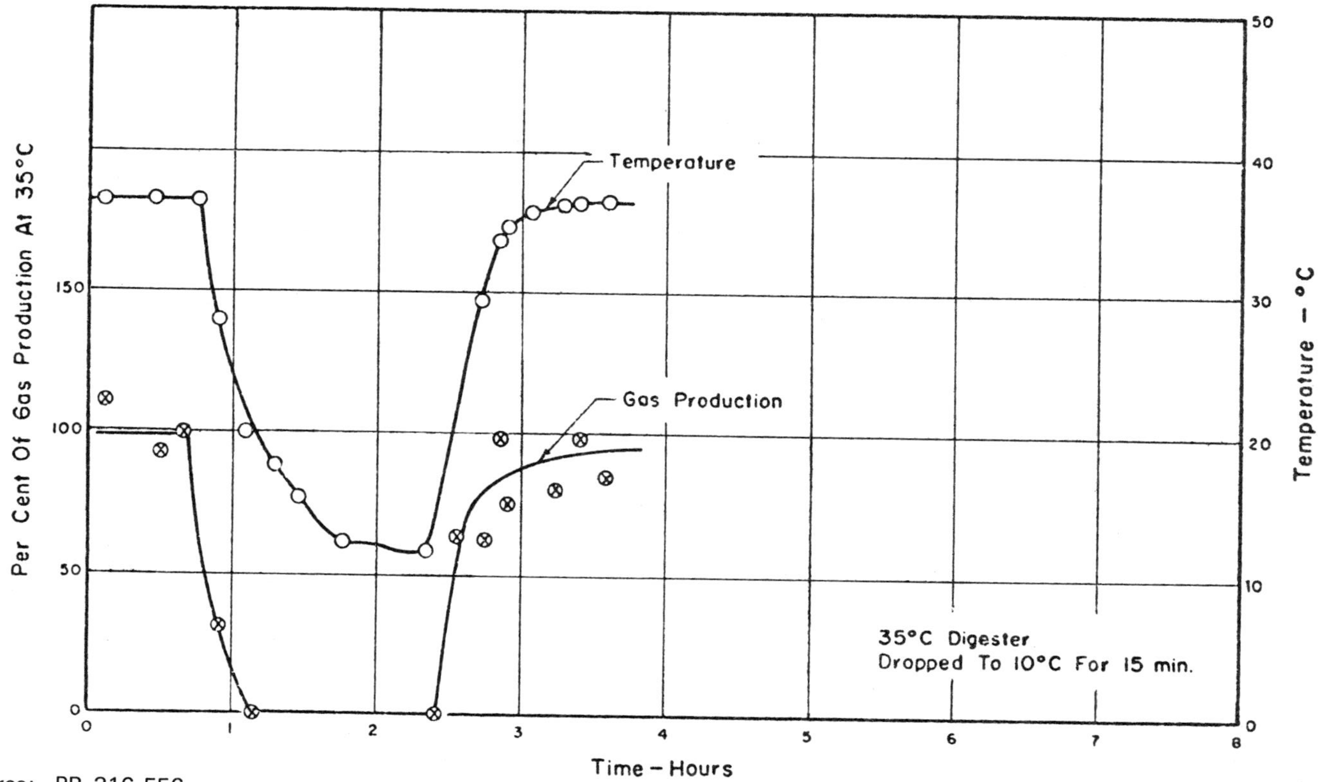

Source: PB 216 556

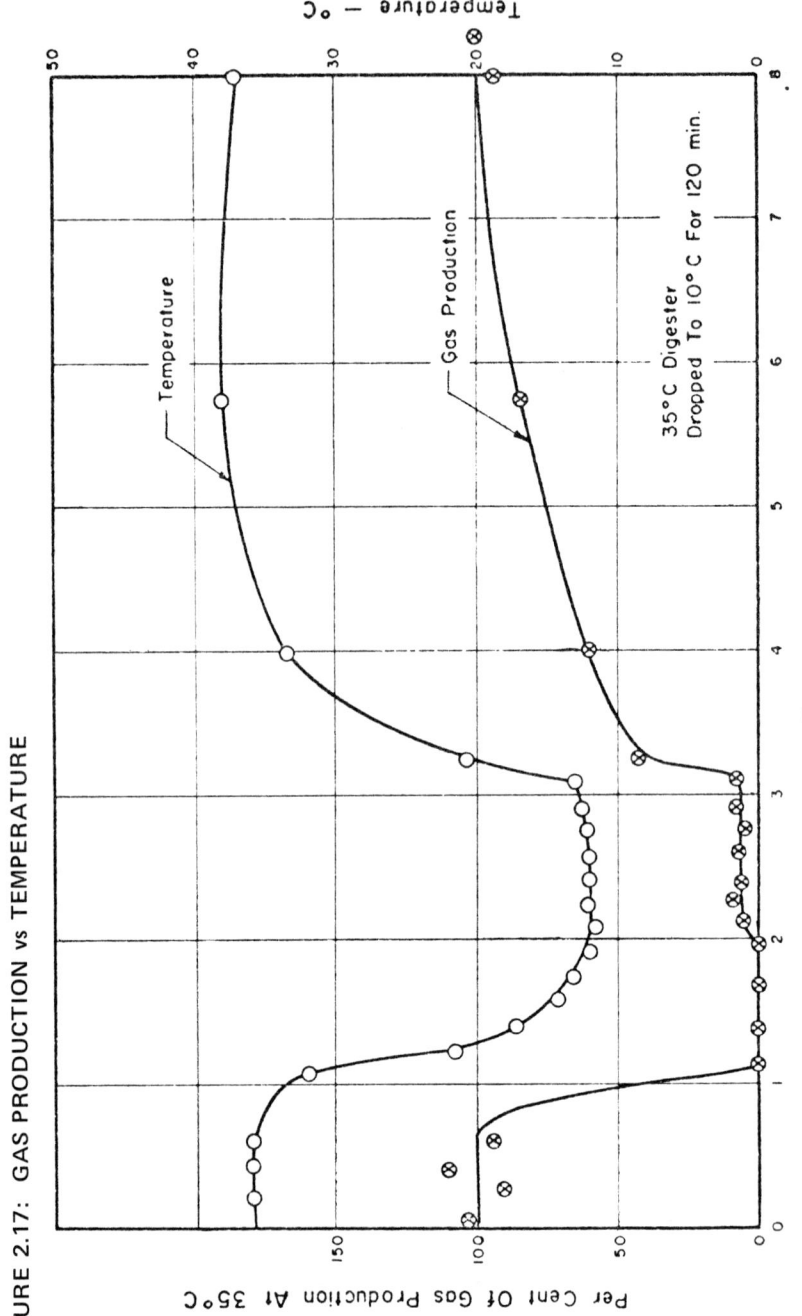

FIGURE 2.17: GAS PRODUCTION vs TEMPERATURE

Source: PB 216 556

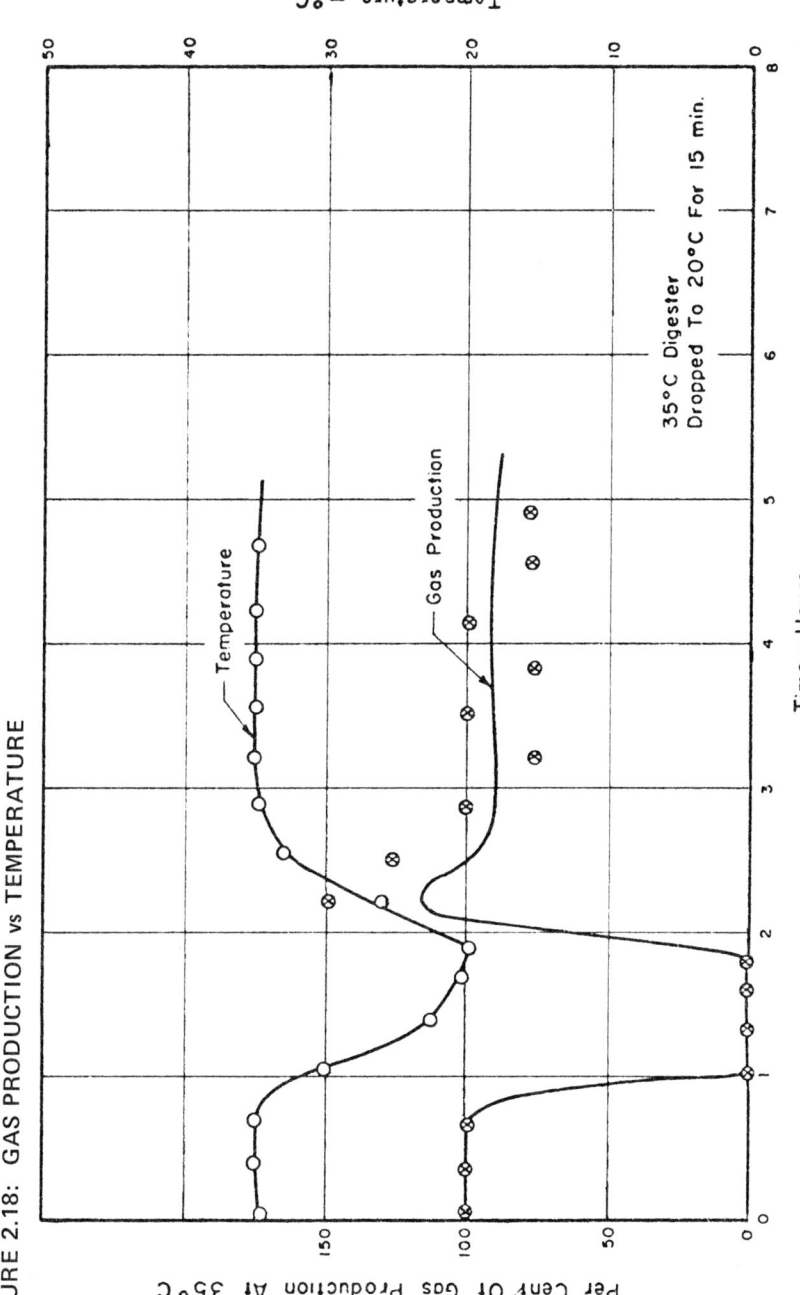

FIGURE 2.18: GAS PRODUCTION vs TEMPERATURE

Source: PB 216 556

60 Energy from Bioconversion of Waste Materials

FIGURE 2.19: GAS PRODUCTION vs TEMPERATURE

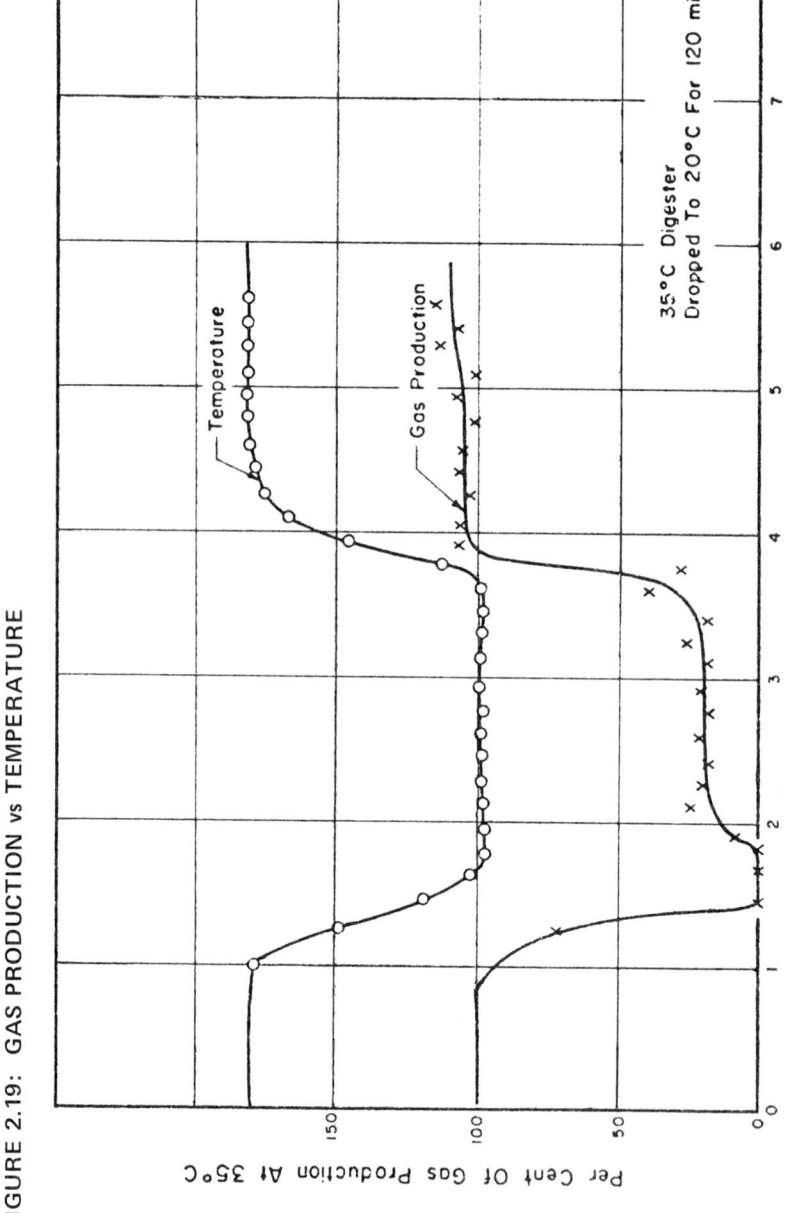

Source: PB 216 556

Bioconversion Mechanisms

Figure 2.17 shows a slight lag in recovery of gas production after maintaining a sludge at 10°C for two hours. It is noted that temperature recovery was also slow in this case due to the type of water bath used during this run. This equipment was modified in later experiments to allow more rapid temperature recovery.

Figures 2.18 and 2.19 show that after equilibrium conditions have been established, about 20% of the initial gas production occurred at 20°C and, again, there was no lag apparent except for the lack of a spike due to CO_2 release.

Figure 2.20 shows no gas production lag for a sludge held at 10°C for two hours and then raised to 50°C. In conclusion to this portion of the experiment, the methane-forming organisms appear to be very adaptable to frequent temperature changes with no adverse effects resulting from temperature drops within the ranges studied.

It was desirable in the course of this study to confirm the effect of temperature on both the acid-formation rate and gas-production rate. To accomplish this, an actively digesting sample of sludge was taken from the primary digester at the Champaign-Urbana Sewage Treatment Plant. This insured that both groups of microorganisms were functioning well. An increase in volatile acids concentration was simulated by adding sufficient acetate to raise the volatile acids concentration of the sludge to 1,800 mg/l, followed by 10% by volume of primary sludge to insure adequate substrate for the acid-forming organisms. The sludge was then divided into two portions and placed in flasks which had been purged of oxygen by flushing with nitrogen and carbon dioxide gas.

The contents of one flask were maintained at 35°C and the second at 45°C. Both flasks were continually mixed and once a day 1/20 of the contents was withdrawn and replaced by an equal amount of raw sludge. The raw sludge feed simulated normal digester operation and provided a food source for the acid-forming microorganisms, and the volatile acids provided a food source for the methane-forming organisms. Thus, by comparing the volatile acids concentration and gas production in the 45°C digester with that in the 35°C digester, the effect of a temperature increase on the relative rates of activity of the two groups of organisms, the acid-formers and the methane-formers, could be determined.

If the activity of both groups of microorganisms was dependent to the same degree of temperature, the volatile acids concentration in both the 35° and 45°C digesters should be equal, because the increased volatile acid utilization would be matched by a corresponding increased volatile acid formation. If the volatile acids concentration was lower in the 45°C digester than in the 35°C digester, this would indicate that the activity of the methane-formers increased at a greater rate with temperature than the acid-formers.

Figure 2.21 reveals a very significant comparison. It shows the beneficial effect of increased temperature on a digester in which the methane-formers are not inhibited by anything but temperature; that is, there are no adverse environmental or physiological conditions restricting the rate of activity of the methane-formers except temperature. In less than 30 hours, the volatile acid concentration in the 45°C digester had dropped to the original level of 300 mg/l. However, in the 35°C digester the volatile acids concentration had dropped only to about 1,400 mg/l.

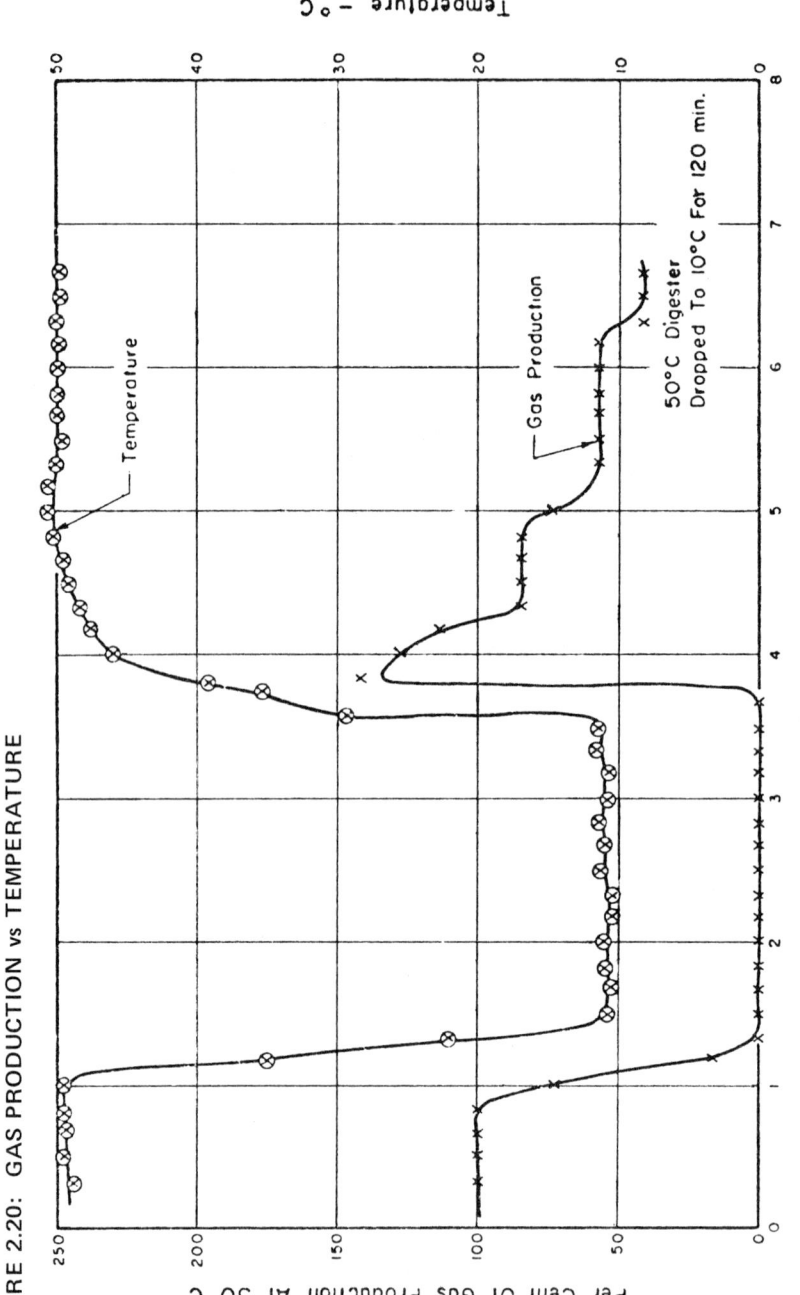

FIGURE 2.20: GAS PRODUCTION vs TEMPERATURE

Source: PB 216 556

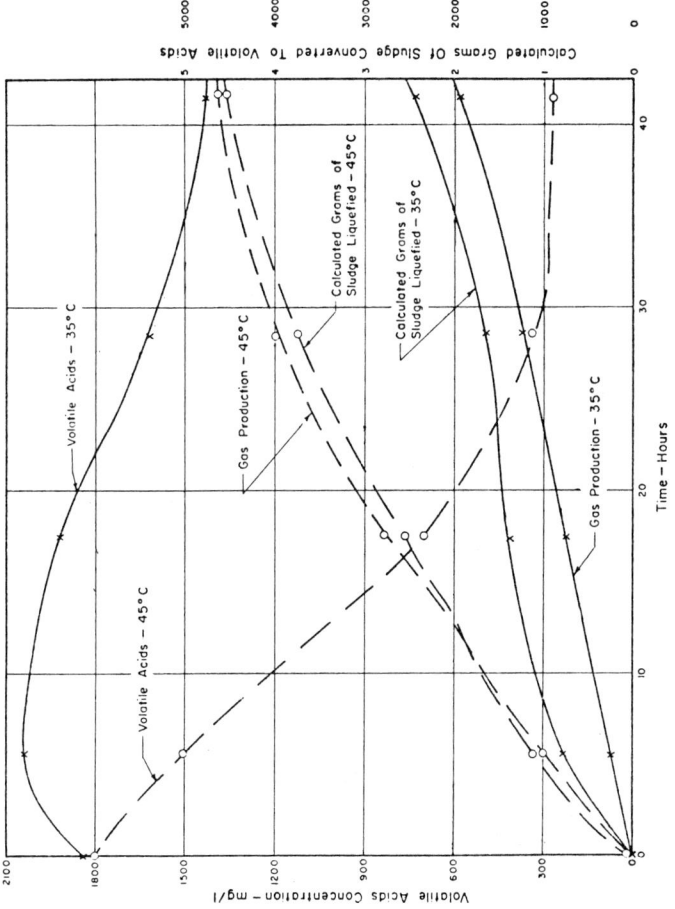

FIGURE 2.21: COMPARATIVE DIGESTER PERFORMANCE AT 35°C AND 45°C

Source: PB 216 556

It would be expected that the increase in temperature would not only increase the rate of activity of the methane-formers, but also the rate of activity of the acid-formers. This is borne out by calculation on the data from each 2-liter digester. At 38.5 hours, 4,000 ml of gas was produced by the 45°C digester and the volatile acids had decreased by 2.92 grams in the 2-liter digester. Assuming 600 ml of gas is produced per gram of volatile acids destroyed, then 4,000 ml of gas represents the destruction of 6.67 grams of volatile acids.

Since the amount of volatile acids in the digester was only reduced by 2.92 grams, then the remaining 3.75 grams of volatile acids must have been contributed by the acid-formers during this time. By similar calculations, it is found that only 1.64 grams of volatile acids were contributed by the acid-formers in the 35°C digester in the same time.

It is concluded from Figure 2.21 that while an increase in temperature does increase the activity of both the methane-forming organisms and the acid-formers, the methane-formers increase their activity at a considerably greater rate per degree. Thus, not only can the methane-formers consume the additional volatile acids contributed by the acid-formers at higher temperatures, but in addition, they are able to reduce the pool of volatile acids which is initially present. This explains the rapid decrease in volatile acids in the 45°C digester as compared to the less rapid decrease in the 35°C digester.

Engineering Significance: Figure 2.22 is hypothesized as a characteristic plot of the temperature dependence of the acid-forming and methane-forming microorganisms in a favorable anaerobic digestion environment. Above a certain temperature, X, the methane-formers are capable of consuming volatile acids at a rate greater than the rate at which they are supplied by the acid-formers. Therefore, the volatile acids concentration remains low. Below this temperature, X, the rate of activity of the methane-formers is lower than the acid-formers. The net result is that volatile acids start to accumulate in the digester.

Experience seems to indicate that the temperature, X, at which the rates are equal is approximately 30° to 35°C. There appears to be only a minor advantage to be gained by maintaining the temperature of a sewage sludge digester above this range, since the rate of liquefaction is not appreciably increased, according to Golueke (40) and Heukelekian and Kaplovsky (41). The results, as shown in Figure 2.21, indicate a definite rate increase in acid-formation but not as much as the rate increase in the methane formation. Above temperature X, acid formation is the lower and thus the rate-limiting step in the two-step digestion process, and the increased potential capacity of the methane-formers cannot be exploited because they have only a limited food supply.

In the anaerobic digestion of a soluble industrial waste, the rate of acid formation may be appreciably greater than with a complex solid, like sewage sludge. In such a case, advantage could be taken of the markedly greater rate of methane formation at the 40° to 45°C level as compared to the 35°C commonly used. This would result in a much smaller digester.

Premixing of digested sludge with raw sludge before pumping the mixture into the digester appears to stop methane production if the temperature of the mixture would be lowered to less than 20°C. However, no retardation of gas production results when the temperature is restored to normal inside the digester.

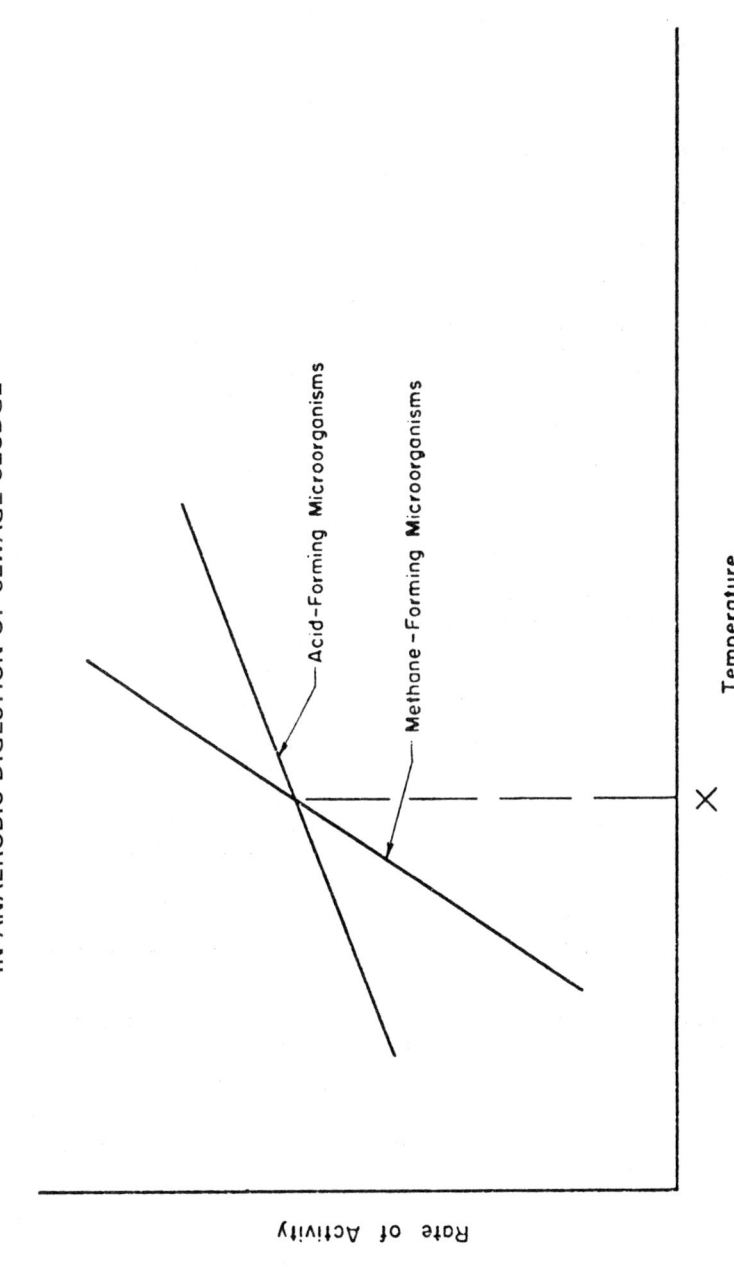

FIGURE 2.22: TEMPERATURE DEPENDENCE OF ACID-FORMING AND METHANE-FORMING MICROORGANISMS IN ANAEROBIC DIGESTION OF SEWAGE SLUDGE

Source: PB 216 556

On this basis, the flow diagram proposed in Figure 2.12 appears to be feasible with respect to the methane-forming organisms in that they are able to adapt to the changing temperatures. The adsorptive concentration of the pollutant would be critical and require considerable study.

REFERENCES

(1) Hungate, R.E., *The Rumen and Its Microbes*, New York: Academic Press, Inc., 1966.
(2) Hungate, R.E., "Studies on Cellulose Fermentation. I. The Culture and Physiology of an Anaerobic Cellulose-Digesting Bacterium," *Journal of Bacteriology*, 48:499-513, 1944.
(3) Baker, F., "Direct Microscopical Observations upon the Rumen Population of the Ox. I. Qualitative Characteristics of the Rumen Population," *Annals of Applied Biology*, 30:230-239, 1943.
(4) Hungate, R.E., "Studies on Cellulose Fermentation. III. The Culture and Isolation of Cellulose-Decomposing Bacteria from the Rumen of Cattle," *Journal of Bacteriology*, 53:631-645, 1947.
(5) Enebo, L., "Symbiosis in Thermophilic Cellulose Fermentation," *Nature*, 163:805, 1949.
(6) Hungate, R.E., "The Anaerobic Mesophilic Cellulolytic Bacteria," *Bacteriological Reviews*, 14:1-49, 1950.
(7) McBee, R.H., "The Anaerobic Thermophilic Cellulolytic Bacteria," *Bacteriological Reviews*, 14:51-63, 1950.
(8) Hammerstrom, R.A., Claus, K.D., Coghlan, J.W. and McBee, R.H., "The Constitutive Nature of Bacterial Cellulases," *Archives of Biochemistry and Biophysics*, 56:123-129, 1955.
(9) Kitts, W.D. and Underkofler, L.A., "Hydrolytic Products of Cellulose and the Cellulolytic Enzymes," *Journal of Agricultural Food Chemistry*, 2:639-645, 1954.
(10) Halliwell, G., "Cellulolytic Preparations from Microorganisms of the Rumen and from *Myrothecium verrucaria*," *Journal of General Microbiology*, 17:166-183, 1957.
(11) Goto, M. and Okabe, N., "Cellulolytic Activity of Phytopathogenic Bacteria," *Nature*, 182:1516, 1958.
(12) Hungate, R.E., Phillips, G.D., McGregor, A., Hungate, D.P. and Buechner, H.K., "Microbial Fermentation in Certain Mammals," *Science*, 130:1192-1194, 1959.
(13) Husain, A. and Kelman, A., "The Role of Pectic and Cellulolytic Enzymes in Pathogenesis by *Pseudomonas solanacearum*," *Phytopathology*, 48:377-386, 1958.
(14) Bryant, M.P. and Robinson, I.M., "Some Nutritional Requirements of the Genus Ruminococcus," *Applied Microbiology*, 9:91-95, 1961.
(15) Bryant, M.P. and Robinson, I.M., "Studies on the Nitrogen Requirements of Some Ruminal Cellulolytic Bacteria," *Applied Microbiology*, 9:96-103, 1961.
(16) Hungate, R.E., "Symposium: Selected Topics in Microbial Ecology," *Bacteriological Reviews*, 24:353-364, 1960.
(17) Kistner, A., "An Improved Method for Viable Counts of Bacteria of the Bovine Rumen Which Ferment Carbohydrates," *Journal of General Microbiology*, 23:565-576, 1960.
(18) Storvick, W.O. and King, K.W., "The Complexity and Mode of Action of *Cellvibrio gilvus*," *Journal of Biological Chemistry*, 235:303-307, 1960.
(19) Moscatelli, E.A., Ham, E.A. and Rickes, E.L., "Enzymatic Properties of a β-Glucanase from *Bacillus subtilis*," *Journal of Biological Chemistry*, 346:2858-2862, 1961.
(20) Strider, D.L. and Winstead, N.N., "Production of Cell-Wall Dissolving Enzymes by *Cladosporium cucumerinum* in Cucumber Tissue and in Artificial Media," *Phytopathology*, 51:765-768, 1961.
(21) Brooks, M.A., "The Microorganisms of Healthy Insects," *Insect Pathology*, Volume I, ed. E.A. Steinhaus, New York: Academic Press, pp. 215-150.
(22) Hungate, R.E., "Further Experiments on Cellulose Digestion by the Protozoa in the Rumen of Cattle," *Biological Bulletin*, 84:157-163, 1943.
(23) Hungate, R.E., "Polysaccharide Storage and Growth Efficiency in *Ruminococcus albus*," *Journal of Bacteriology*, 86:848-860, 1963.

(24) Hobson, P.N. and Smith, W., "Continuous Culture of Rumen Bacteria," *Nature*, 200:607-608, 1963.
(25) Hobson, P.N., "Continuous Culture of Rumen Bacteria: Apparatus," *Journal of General Microbiology*, 38:161-166, 1965.
(26) King, K.W. and Vessal, M.I., "Enzymes of the Cellulase Complex," *Cellulases and Their Applications*, Advances in Chemistry Series, 95:7-25, 1969.
(27) Leatherwood, J.M., "Cellulose Complex of Ruminococcus and a New Mechanism for Cellulose Degradation," *Cellulases and Their Applications*, Advances in Chemistry Series, 95:53-59, 1969.
(28) Prins, R.A., van Vugt, F., Hungate, R.E. and van Vorstenbosch, C.J.A.H.V., "A Comparison of Strains of *Eubacterium cellosolvens* from the Rumen," *Antonie van Leeuwenhoek*, 38:153-161, 1972.
(29) Leatherwood, J.M. and Sharma, M.P., "Novel Anaerobic Cellulolytic Bacterium," *Journal of Bacteriology*, 110:751-753, 1972.
(30) Kingsley, V.V. and Hoeniger, J.F.M., "Growth, Structure, and Classification of Selenomonas," *Bacteriological Reviews*, 37:479-521, 1973.
(31) Han, Y.W. and Callihan, C.D., "Cellulose Fermentation: Effect of Substrate Pretreatment on Microbial Growth," *Applied Microbiology*, 27:159-165, 1974.
(32) Hungate, R.E., "The Culture of *Euplodinium neglectum*, with Experiments on the Digestion of Cellulose," *Biological Bulletin*, 83:303-319, 1942.
(33) Coleman, G.S., "The Growth and Metabolism of Rumen Ciliate Protozoa," *Symbiotic Associations*, Thirteenth Symposium, Society for General Microbiology, Cambridge: Cambridge University Press, Inc., 1963, pp. 298-325.
(34) Hungate, R.E., "Studies on Cellulose Fermentation. II. An Anaerobic Cellulose Decomposing Actinomycete, *Micromonospora propionici*, n. sp.," *Journal of Bacteriology*, 51:51-56, 1946.
(35) Reese, E.T., Mandels, M. and Weiss, A.H., "Cellulose as a Novel Energy Source," *Advances in Biochemical Engineering*, ed. T.K. Ghose, A. Fiechter and N. Blakebrough, Berlin: Springer-Verlag, 1972, pp. 181-200.
(36) Festenstein, G.N., "Cellulolytic Enzymes from Sheep-Rumen Liquor Microorganisms," *Biochemical Journal*, 69:562-567, 1958.
(37) Halliwell, G., "Cellulolysis by Rumen Microorganisms," *Journal of General Microbiology*, 17:153-165, 1957.
(38) Matthijssen, C., "Studies on the Cellulolytic Enzyme System of *Corynebacterium fimi*," *Dissertation Abstracts*, 17:2141, 1957.
(39) Balch, D.A., "An Estimate of the Weights of Volatile Fatty Acids Produced in the Rumen of Lactating Cows on a Diet of Hay and Concentrates," *British Journal of Nutrition*, 12:18-24, 1958.
(40) Golueke, C.G., "Temperature Effects on Anaerobic Digestion of Raw Sewage Sludge," *Sewage and Industrial Wastes*, 30:1225, 1958.
(41) Heukelekian, H. and Kaplovsky, A.J., "The Effect of Change of Temperature on Thermophilic Digestion," *Sewage Works Journal*, 20:806, 1948.
(42) Garber, W.F., "Plant-Scale Studies of Thermophilic Digestion at Los Angeles," *Sewage and Industrial Wastes*, 26:1202, 1954.

BIOCONVERSION OF SOLID WASTE AND SEWAGE SLUDGE

The sources of the material in this chapter were the following reports:
PB 231 149, PB 231 176, PB 235 468, PB 238 068, N75-25292, PB 245 083, PB 238 563, PB 245 083, PB 220 821 and U.S. Patent 3,640,846. For a complete bibliography, see page 222.

UNIVERSITY OF ILLINOIS STUDIES

Anaerobic Processing Costs

At the Bioconversion Energy Research Conference held at the University of Massachusetts in June 1973, J.T. Pfeffer of the University of Illinois described a study done for the Environmental Protection Agency. The study was initiated in the Fall of 1967 and completed in January 1973. The objective of the work was to evaluate the potential for processing organic solid waste for the purpose of producing a fuel gas, i.e., methane. The study elucidated the potential operating problems and environmental conditions required for the least cost production of this product.

The substrate for these studies consisted of domestic refuse that was obtained from a residential area in Cincinnati. The refuse was shredded at the EPA's Center Hill Research facility. Nominal particle size was in the range of one to two inches. The refuse was found to be deficient in nitrogen and phosphorus. Therefore, a decision was made to investigate the advantages of the use of sewage sludge as an additive to the process. The addition of a small quantity of raw sewage sludge from the Urbana-Champaign treatment plant did significantly improve the digestion. However, the substrate was still deficient in nitrogen and phosphorus and supplemental nutrients were required.

The quantity of sewage sludge generated by a population is not adequate to meet the nutritional needs for processing of the organic refuse from this same population. The substrate chemical composition was such that adequate buffer did not develop in the digestive process. It was impossible to maintain the reactor pH above 6.6 and it was necessary to add caustic to the reactors to maintain an acceptable pH level. After it was established that the refuse could be converted to methane with the proper addition of nutrients and caustic, the study investigated the effect of temperature and retention time on this degree of conversion. The results suggest that the system can be effectively used. Gas production is shown in Table 3.1.

TABLE 3.1: GAS PRODUCTION*

Temperature (°F)	Retention Time (days)					
	4	8	10	15	20	30
95.0	1.40	2.30	2.58	2.86	2.95	3.05
104.0	2.33	2.98	3.20	3.60	3.85	4.10
107.8	2.59	3.14	3.33	3.74	3.95	4.20
109.4	2.58	3.09	3.26	3.67	3.82	4.09
113.0	2.08	2.55	2.70	3.05	3.14	3.36
118.4	2.64	3.14	3.33	3.57	3.82	4.06
127.4	3.33	3.88	4.07	4.29	4.51	4.70
132.8	3.74	4.20	4.35	4.54	4.75	4.88
140.0	4.23	4.49	4.57	4.79	4.93	4.96

*Cu ft/lb dry solid

Source: PB 231 149

As would be expected, the quantity of gas increases with increasing temperature and time. The discontinuity on this data at a temperature of approximately 110° to 113°F indicates the transition from a population of mesophilic microorganisms to a population of thermophilic microorganisms.

An increase in retention time in the mesophilic temperature range provided a significant increase in gas production even at retention times of 20 to 30 days. However, in the thermophilic temperature range the increase in gas yields with increasing retention times was almost insignificant beyond about 10 days' retention time. Because of the high temperature and related microorganism activity, the biodegradable fraction of the refuse was rapidly converted to methane gas. Therefore, short hydraulic retention times would be appropriate when operating in the thermophilic temperature range. The gas composition in the mesophilic and thermophilic range is shown in Table 3.2.

TABLE 3.2: GAS COMPOSITION AT 35° AND 60°C

Reactor Number	35°C			60°C		
	Retention Time (day)	%CH$_4$	%CO$_2$	Retention Time (day)	%CH$_4$	%CO$_2$
1	4	69.7	30.3	3	53.5	46.5
2	4	69.8	30.2	4	55.0	45.0
3	6	64.3	35.7	6	51.9	48.1
4	8	58.6	41.4	8	50.2	49.8
5	10	57.2	42.8	10	52.3	47.7
6	15	53.8	46.2	15	53.5	46.5
7	20	53.4	46.6	20	49.1	50.9
8	30	53.8	46.2	30	53.7	46.3

Source: PB 231 149

The methane content decreases with increasing temperature and retention time. This variation is caused by the increased washout of carbon dioxide in a liquid phase at short retention time and a decrease in solubility of carbon dioxide at high temperatures. The gas composition is very pH and alkalinity dependent. The highest methane content was obtained at the 35°C temperature and short retention time. Under these conditions the methane accounts for approximately 70% of the gas. However, in the thermophilic range the methane content was substantially lower, ranging from less than 50 to about 55%. Thermophilic operation would be expected to yield to higher gas cleanup cost than mesophilic.

Measurement of the volatile solids destruction was very difficult because of the nature of the substrate and the reactor size and characteristics. By establishing a careful solids balance, it was found that the dry gas production was 10.92 standard cubic feet per pound of volatile solids destroyed. This agrees well with the theoretical value of 13.3 standard cubic feet per pound of cellulose destroyed when considering the carbon dioxide in the liquid phase. The data in Table 3.3 were calculated from gas production data using a value of 11.0 standard cubic feet per pound of solids destroyed. As can be seen from this table, volatile solids reduction range from a low of 17.4% to a high of 61.8%.

TABLE 3.3: VOLATILE SOLIDS DESTRUCTION, %

Temperature (°F)	Retention Time (days)					
	4	8	10	15	20	30
95.0	17.4	28.6	32.1	36.6	36.7	38.0
104.0	29.0	37.1	39.6	44.8	47.9	51.0
107.8	32.2	39.1	41.5	46.6	49.2	52.3
109.4	32.1	38.5	40.6	45.7	47.6	50.9
113.0	25.9	31.7	33.6	38.0	39.1	41.8
118.4	32.9	39.1	41.5	44.4	47.6	50.6
127.4	41.5	48.3	50.7	53.4	56.2	58.5
132.8	46.6	52.3	54.2	56.5	59.1	60.8
140.0	52.7	55.9	56.9	59.6	61.4	61.8

Source: PB 231 149

The data reported are based upon a dry solids feed to the system. The moisture content of the refuse is not considered in this analysis. The solids consist of the inorganic as well as the organic fractions. As the microorganisms convert the organic solids to gas, the pounds of solids in the system are reduced. Therefore, the concentration of the solids in the reactor is always less than the concentration being fed to the reactor. Depending upon the degree of solids destruction, this concentration may be as low as 50% of the feed slurry solids concentration. The characteristics of the process should be evaluated based upon the actual solids in the reactor rather than those being fed to the reactor.

In order to evaluate the economic potential of the system, the cost associated with the anaerobic processing and solids dewatering were evaluated. The capital costs were based upon data published by Robert Smith of the federal EPA and were adjusted to January 1973 costs. The capital costs are amortized at a 6% interest rate and a 25-year life. These capital costs include the reactor and the vacuum filtration system. The operating costs associated with the process include mixing power costs for the reactor, heating costs, and power costs for the vacuum filtration system. Mixing costs are based on a power requirement of 0.185 hp/1,000 cu ft at a four percent reactor slurry concentration. The mixing power required is increased linearly with increasing solids concentration. The mixing requirements are considerably less than those required for mass transfer considerations. The degree of mixing is based upon a requirement to keep the reactor content relatively homogeneous.

Heating costs include the heat required to elevate the temperature of the feed slurry, the heat loss through the reactor, and the heat required to vaporize the water that is lost with the gas. As the temperature increases, the partial pressure of water vapor significantly increases, accounting for a substantial heat loss. The analysis assumes no heat recovery.

The dewatering costs were calculated from the required filter area and a power cost of 0.125 hp/ft^2 of filter area. The filter area requirements are based on a filter loading rate determined from laboratory test leaf procedures. The filter loading with this residue is significantly greater than that commonly reported for municipal digester sludge. However, these loadings appear to be reasonable from the limited information currently available. No chemical addition was used and it is conceivable that the filter loading and the moisture of the cake can be significantly changed by the use of chemical treatment, i.e., polymer addition.

The optimum mesophilic temperature was found to be 108°F and the optimum thermophilic temperature was 140°F. The laboratory system was not operated beyond 140°F. Therefore, it is conceivable that slightly higher temperatures may yet improve the gas production. The data in Table 3.4 show the costs associated in operating a system in the mesophilic and thermophilic ranges.

TABLE 3.4: ANAEROBIC PROCESSING COSTS—$/TON OF DRY SOLIDS

Cost Item	Feed Slurry Concentration (% Solids)				
	3	6	10	15	20
	Temperature, 108°F; θ = 15 days				
Capital-Total	11.09	4.15	2.41	1.62	1.22
Digesters	5.80	2.96	1.80	1.23	0.93
Vacuum filter	5.29	1.19	0.61	0.39	0.29
Operating-Total	3.17	1.52	1.01	0.79	0.66
Mixing	0.26	0.26	0.26	0.26	0.26
Heating	1.98	1.06	0.65	0.47	0.36
Dewatering	0.93	0.20	0.10	0.06	0.04
Total Cost	14.26	5.67	3.42	2.41	1.88
	Temperature, 140°F; θ = 8 days				
Capital-Total	9.49	2.88	1.62	1.07	0.82
Digesters	3.15	1.60	0.99	0.68	0.53
Vacuum filter	6.34	1.28	0.63	0.39	0.29
Operating-Total	4.27	2.00	1.28	0.92	0.75
Mixing	0.13	0.13	0.13	0.13	0.13
Heating	3.04	1.66	1.05	0.73	0.57
Dewatering	1.10	0.21	0.10	0.06	0.05
Total Cost	13.76	4.88	2.90	1.99	1.57

Source: PB 231 149

As can be seen from this table, the reactor costs account for a major portion of the production costs in both the mesophilic and thermophilic temperature ranges. The reactor costs relate specifically to the retention time (θ) and the feed slurry concentration when considering a fixed loading rate. In this table calculations are based upon 100 tons per day of dry refuse. Certainly the more concentrated the feed slurry, the smaller the reactor costs.

The mixing requirements relate to the reactor volume and the slurry concentration. The manner in which the mixing costs were calculated show this cost to be constant with increasing feed slurry concentration. This is simply because that for a 100 ton per day plant, a twofold increase in the feed slurry concentration reduces the reactor volume by a factor of two. Therefore, the power input per unit volume is doubled because of the increased slurry concentration. However, the total power required is the same because the total volume is only one-half that of the more dilute feed slurry concentration. Both the reactor costs and mixing costs reflect the retention time difference in the mesophilic and thermophilic ranges.

The heating requirements are significant in both temperature ranges. However, the thermophilic range requires substantially more heat than the mesophilic range. Since no heat recovery has been included in these calculations, it is possible to reduce these costs significantly by instituting a heat recovery program. In this way, both the heat from the reactor effluent and the off gases could be recovered.

The dewatering costs are significantly reduced with increasing slurry concentration. This results primarily from a marked increase in filter loading with a more concentrated reactor slurry. The higher dewatering costs associated with thermophilic temperature results primarily from the technique used in calculating these costs. Because of a higher destruction of volatile solids at the thermophilic range, the reactor slurry is more dilute. Therefore, the filter loading is reduced with this more dilute slurry. The decrease in the total pounds of solids for dewatering at the thermophilic temperature does not offset this calculated reduction. The validity of this difference may be questioned but the dewatering costs are sufficiently accurate for comparative purposes.

From this analysis it could be suggested that the system is relatively insensitive to mixing costs and dewatering costs. The primary cost elements seem to be the reactor capital requirements and the heating requirement. If it is found that the power required for mixing does not linearly increase with the reactor slurry concentration, it is conceivable that the mixing could play a more important role in the production costs. However, it is unlikely that the mixing power will increase by more than a factor of two. The power input for the feed slurry concentration of 15% is 0.46 hp/1,000 cu ft or 0.4 hp/1,000 cu ft for the mesophilic and thermophilic temperature, respectively. If this power input is doubled, it would approach 1 hp/1,000 cu ft which is a substantial mixing power input.

Calculations of the cost of producing gas show that for a 15% feed slurry concentration, the cost is 13.7 cents per 1,000 cubic feet at a temperature of 108°F. For the thermophilic temperature the minimum cost was 9.6 cents per 1,000 cubic feet. Increase in the feed slurry concentration to 20% produces a cost of 11.1 and 7.8 cents per 1,000 cubic feet for the mesophilic and thermophilic range, respectively. If this gas can be marketed at a value of $0.75 per million Btu, a return of $2.5 to $3.0 per ton of dry refuse can be expected. Additional cost benefits can also be included as credits. This would include the processing of approximately 2.9 tons of dry sewage solids per 100 tons of refuse. It would also include a reduction in the volume of landfill required by approximately one-third that for unprocessed refuse. If the preprocessing of shredding and separation can be accomplished at a cost between $2 and $3 per ton, this system will likely produce a net income.

A detailed description of the experimental procedure used and the results obtained can be found in the NTIS report, PB 231 176. See p 222 for the complete bibliographic entry.

Fermentor Residue Dewatering

A major cost item associated with anaerobic processing is the cost of dewatering the fermentor residue and the disposal of the residue from the system. J.T. Pfeffer and J.C. Liebman, in further work, devised an experimental program to advance the state of the art knowledge of these operations to accurately evaluate the costs. The results of these experimental evaluations are as follows.

Fermentation System: A fermentor consisting of a mixed 100 gallon reactor was used. During the period covered, July 1973 to July 1974, refuse from Champaign, Illinois, Madison, Wisconsin, and St. Louis, Missouri was used as feed for this system. Operation of the process was routine. Equipment malfunction did not have any pronounced effect on the fermentation system once a temperature control problem was corrected. Operating at 64°C has a significant adverse effect on the methane fermenting microorganisms developed at an operating temperature of 60°C.

Gas production reflects the characteristics of the refuse being used as the substrate. The gas production was 11.0 and 12.0 scf/lb of volatile solid destroyed for the Champaign and Madison refuse respectively. The gas produced by the higher quality Champaign refuse was 6.28 scf/lb of volatile solids added. The Madison refuse was of lower quality and produced only 5.22 scf/lb of volatile solids added.

Rheological Properties of Reactor Slurry: Because of the character of the slurry, it was not possible to measure the viscosity of the slurry when the solids exceeded 25 g/l. The filtrate viscosity was found to be 1.5 cp when no chemicals were added for conditioning the sludge prior to dewatering. The viscosity and yield stress of the reactor slurry increase approximately exponentially with solids concentration over the range of solids tested. The reactor slurry exhibited a higher viscosity than digested sewage sludge for a given solids concentration. Both slurries used in these tests exhibit Bingham plastic behavior.

Filterability of Reactor Effluent: Filter test leaf studies were conducted on the reactor effluent to determine the dewatering characteristics of the residue. The filter yield is a function of both the feed solids concentration and cake solids desired. With a reactor slurry of 5 to 6% total solids, filter yields between 30 and 40 lb/ft^2/hr can be obtained if a cake solids of 20% is acceptable. A cake solids of 25% can be obtained with an associated filter yield of approximately 20 lb/ft^2/hr. Higher cake solids greatly reduce the filter yield when chemical conditioning is not employed as was the case for the above data. Specific resistance measurements were conducted on digested sewage sludge and reactor slurry.

The specific resistance of digested sewage sludge was 1.2×10^{10} sec^2/g. The unconditioned reactor slurry contained a low quantity of fine solids. When these fine solids were present in quantity, as was the case with the Madison refuse, specific resistance could not be measured because these solids clogged the filter paper almost at once. When this slurry was conditioned with a polymer, the specific resistance was 7.5×10^7 sec^2/g at an initial solids concentration of 30.8 g/l and 2.2×10^6 sec^2/g at 78.8 g/l of solids.

Conditioning of the slurry prior to filtration greatly improved the filtration results. Cake solids in excess of 30% were obtained at filter yields of 20 lb/ft^2/hr with a feed solids concentration of about 5%. A polymer dose of 20 lb/ton of dry solids was required under these test conditions. Solids capture with polymer addition was in excess of 95%.

Centrifugation of Reactor Effluent: A 14-inch diameter solid bowl basket centrifuge was used to evaluate the dewatering properties of this reactor residue. The maximum solids capture was obtained at 2,350 rpm which produced a centrifugal force of 1,050 *g*. Higher speeds did not improve solids capture. Chemical conditioning was not used in these tests. The solids capture could be improved if chemical conditioning were employed.

The residue had excellent dewatering properties at the higher speeds. With a speed of 1,050 rpm (200 *g*), the cake was only 13% solids. At 2,350 rpm (1,050 *g*), the cake solids increased to 27%. At the maximum speed of the centrifuge, 3,400 rpm (2,200 *g*), the solids content of the cake reached 32.5%. The results of the tests indicate that centrifuging may be the preferred dewatering technique when considering the final cake solids.

Settleability of Solids in the Reactor Effluent: The reactor effluent has very poor settling characteristics when the solids concentration is greater than 3%. In order to thicken the solids to 4% solids, the solids flux should not exceed 2.1 lb/ft^2/hr. Thickening to 5% solids reduces this flux to only 0.75 lb/ft^2/hr. Since the reactor would be operating with solids concentration of 5% or greater, there is no practical application for thickening in this process.

Leachate Potential of Dewatered Reactor Residue: A laboratory scale lysimeter was charged with filter cake to evaluate the gross pollution potential of this material. The COD of the leachate varied from 1,000 to 10,000 mg/l while the total dissolved solids varied from 3,500 to 11,000 mg/l. The dissolved solids were always 50% or more inorganic. The peak values for COD and dissolved solids lasted for only a short time indicating that some of the contamination resulted from the purging of the liquid associated with the cake.

Calorific Value of the Reactor Residue: Bomb calorimeter studies of the dry filter cake from the reactor were conducted to determine the heating value of the refuse. The dry residue contained 7,334 Btu/lb (high heating value). The residue was 75.3% volatile. Therefore, the heating value was 9,734 Btu/lb of volatile solids.

Computer Simulation Results: The computer runs show that the overall process of producing methane from refuse is economically sound. The economics are most sensitive to the credit allowed for refuse disposal. The cost applied to the "front-end" and the incinerator costs are questionable. While wide variations in costs associated with the various unit processes will affect the net income, reasonable estimates of these costs do not make the process unprofitable. The optimum detention time is a function of gas value. As the gas value increases, the optimum retention time shifts from short periods to longer periods. At 60°C, the minimum retention time evaluated (5 days) produces maximum net gain when gas is priced at $1,000/Mcf. Within the practical limits imposed by materials handling constraints (mixing, etc.), the higher the feed solids concentration, the greater the net gain ($/hr) and the net energy recovery. Operating at 60°C produces a greater net energy recovery as well as a greater net gain ($/hr) than the mesophilic temperature of 40°C.

A detailed description of this work is provided in NTIS report, PB 235 468. See p 222 for the complete bibliographic entry.

DYNATECH STUDY

At the Bioconversion Energy Research Conference of June 1973, D.L. Wise described work being done by Dynatech. For many years scientists and engineers at Consolidated Natural Gas Service Company, Inc. have studied virtually all techniques for supplementing the natural gas resources of the United States. As part of this continuing effort, Consolidated has had carried out, at Dynatech R/D Company, laboratory experiments on the production of transmission line fuel gas by the anaerobic digestion of the large (50% by weight) organic fraction of solid waste. This process has broad national application and in no way is limited to a particular area of the country. Preliminary estimates indicate that larger plants in metropolitan areas will produce fuel gas at prices competitive with natural gas and that use of the process in smaller municipalities may produce gas competitive with present supplemental gas.

Anaerobic digestion of the organic fraction of solid waste offers several unique features of total environmental control. It upgrades the lowest quality fraction in solid waste to the least polluting combustible hydrocarbon, methane. The utilization process itself requires no air pollution control devices such as would be needed for incineration of solid waste to produce direct heating such as steam. The product gas is storable which leads to its application for supplemental energy requirements as well as permitting optimum operation of the process. Further, the process is carried out in an aqueous medium in which it is very desirable to recycle the process water.

Thus a solid waste problem does not become a major water pollution problem. As a further example of the total pollution control aspect of this system, it is of interest to note that the nutrient requirements of the microorganisms can be satisfied by the utilization of sewage sludge, thus eliminating the pressing ultimate disposal problem of a municipal waste treatment plant. The overall anaerobic digestion process itself is familiar to those personnel in a municipal government responsible for waste treatment and no major retraining program will be required to introduce this process into the community. As a long range program for reducing the costly collection of solid waste, it is possible to visualize a system to prepare solid waste in suitable form at the home and transport all waste material through sewage lines to a centralized beneficial waste utilization facility. The highlights of the experimental research program on fuel gas from solid waste are as follows. Initial work centered on demonstrating the practicality of fuel gas production from solid waste. Laboratory digesters were acclimated to 100% solid waste as a feedstock. These digesters initially operated on sewage sludge and were converted to consuming waste. A tight operating and analysis system was developed. Eight 50-liter digesters at mesophilic conditions and four 50-liter digesters at thermophilic conditions were operated. Gas production was approximately 10 ft^3/lb of feed using from 15 to 30 days solids retention times. Product gas composition was 70% methane for the sewage sludge feedstock evaluated and approached 60% methane for solid waste feedstock.

A number of features of laboratory digester operation related to integrating this fuel gas production step into the overall solid waste system. For example, effluent from the laboratory digesters was used to prepare the shredded solid waste as will be done on a large scale in which recycle of effluent water as well as microorganisms is essential. Also, the laboratory method used for preparing solid waste achieved substantial particle size reduction which appeared to enhance enzymatic breakdown of the cellulosic material.

Engineering analysis and system optimization have been carried out based on a process handling 1,000 tons per day of solid waste. It has been assumed that 50% of this input stream is biodegradable and that this biodegradable portion is largely cellulosic. The design is based on the so-called anaerobic contact process, whereby the microorganism containing effluent from the digester is settled out and then recycled. The digester is assumed to operate at 35°C (100°F) and heat is assumed to be conserved by effluent water recycle to the wet separation and preparation portion of the overall process; in this way only a relatively small portion of the total heat is required for makeup.

On this process design basis a computer program was written and the minimum total cost, including capital and operating expenses, was obtained. The optimum total cost system was obtained using reliable engineering values for the required constants, such as microorganism growth rate data, mixing power requirements, settling velocities, and basic construction and utility cost data. This information was obtained from the literature from experiments carried out on sewage sludge. With this reasonably conservative set of engineering constants, a parameter study was carried out on the computer. That is, using a set of base data for all engineering constants each specific engineering constant was varied in turn over the fullest expected range and the corresponding minimum total cost obtained.

A detailed computer optimization was then carried out based on experimentally determined design parameters. The results are for a conventional mesophilic anaerobic contact process and leave room for innovative and creative engineering improvements. Included in this work is a complete computer sensitivity analysis. In every respect the process appears to be attractive on an economic basis.

One practical operating feature brought out in the computer study was the significance of having the feedstream to the digester as free from nonbiodegradable material as possible. With the microorganism recycle system planned, even percentage points of nonbiodegradable material in the feed will require substantial bleed-out and hence loss of needed methane-producing microorganisms. Thus, investigation of solid waste separation and preparation techniques has received considerable attention. A detailed process description for the 1,000 tpd solid waste to fuel gas facility and a brief description of the computer model results are presented below.

Process Description

Basically, the process can be divided into four different areas of operation:

(1) Municipal Waste Handling Operations—These operations consist of the receiving operation, separation of nondigestibles, and resource recovery operations.

(2) Digestion—That step includes the feeding of nutrients and pH control additives necessary for the satisfactory operation of the digesters.

(3) Gas Treatment—Methane must be scrubbed free from carbon dioxide and dried before it is ready for distribution.

(4) Effluent Disposal—Both solid and liquid effluents from the process must be adequately disposed of.

Detailed descriptions of the four areas of operation defined above follow.

Waste Receiving and Handling: The municipal waste is delivered to the plant site by packer trucks. Each truck is identified and weighed in an automated weighing station, then driven to a delivery area, a tipping floor, where it dumps its load. The refuse delivered is deposited by front-end loaders onto pit conveyors, which carry it to the size reduction equipment.

The size reduction for equipment fulfills two primary functions: it allows for efficient separation of the organic materials from the inorganic nondigestible matter found in municipal waste, and it allows the digesters to solubilize the feed more readily.

Separation systems based on two different principles are currently being developed in this country and show promise of achieving high degrees of separation: these are the so-called dry and wet separation processes. Dry separation processes are under development by a number of organizations, e.g., the U.S. Bureau of Mines, National Center for Resource Recovery and other industrial organizations; wet separation processes are currently being developed by Black Clawson and others concerned with the recovery of fibers from the waste. Both processes provide a stream with a high concentration of digestible matter, relatively free of metals, glass and grit. The dry separation processes offer the advantage of flexibility in selecting the desired water content of the feed to the digesters, whereas wet separation processes necessarily operate at low solids concentrations in aqueous slurries and hence have the disadvantage of requiring a dewatering step whenever a concentrated solids feed to the digesters is desired. In this study, only the dry separation process has been considered.

Both wet and dry separation processes require the use of primary shredders to reduce the maximum particle size in the stream down to 3 to 6 inches. Magnetic separators remove ferrous metals from the process stream and trommel screens, i.e., revolving screens, are employed to remove fine grit and glass which would otherwise be carried along with the digestibles into the digesters. Revolving screens are believed to be superior to vibrating ones for this application because they have less tendency to blind. In the wet system, the stream is then fed into a hydropulper where it is mixed with a large amount of water. There, additional metal is recovered. Fibrous cellulosic materials are finally obtained, as a dilute aqueous suspension, from the stream via a liquid cyclone. If a concentrated cellulosic stream is desired (greater than about 2 to 3%), dewatering of the stream leaving the separator is necessary. In the dry separation system, the oversize stream leaving the screens is conveyed to an air classifier where the organic materials are separated and recovered as the lighter fraction. The denser fraction, consisting largely of metals, glass and heavy organics may be further processed in a resource recovery stream and readied for recycle. The light organic material fraction is then once again shredded in a secondary shredder (to facilitate mixing in the digester) and conveyed pneumatically to a storage silo where it is finally ready for digestion.

Digestion: Before the stream of recovered organic material is fed into the digesters, it is mixed with nutrients and chemicals necessary for the satisfactory operation of the digesters. The nutrients are supplied by raw sewage sludge obtained from a nearby sewage treatment plant. Lime and ferrous salts are added for pH and for hydrogen sulfide control. If a dry separation operation has been used in the process, it may be necessary to add a sufficient amount of an aqueous stream to the feed to the digester in order to allow handling the mixture, i.e., pumping it, with a slurry pump into the digester. The aqueous stream is conveniently supplied by the liquid effluent from the overall process. Under some conditions it may be necessary to add fresh makeup water; that will occur if high water bleed-off rates occur, such as when the solid residue after digestion is disposed of with a high moisture content.

The cellulosic stream from the separators, sewage sludge, chemicals and, when needed, the aqueous stream are metered and combined in a mixing tank where they are blended and fed into the battery of digesters. Each digester consists of a large tank, preferably with a floating cover to maintain the system at constant pressure. These tanks, often circular, are provided with means for stirring the contents continuously. Such mixing may be carried out by sludge recirculation pumps, gas draft-tube mixers, or turbine or propeller mixers. Stirring allows uniform digestion of the material to take place in the digesters. Digestion occurs in two steps: solids are first solubilized by enzymatic action; the soluble products are then digested in a series of reactions finally giving methane and carbon dioxide.

In mesophilic digestion, the process used, the contents of the digesters are maintained at a temperature of about 95° to 100°F. This can be achieved by preheating the aqueous stream in a heat exchanger with steam, or by injecting steam directly into the feed slurry streams. Because of the large digester volumes used, and consequently their small surface-to-volume ratios, minimal insulation is needed to keep heat losses low. The major heat requirement is the sensible heat rise of the feed. In winter, the temperature of the sewage sludge may drop to 50°F and that of the solids even lower. During that period of the year the heating requirement is substantially higher than during the warmer months.

The products of digestion consist of two streams. One stream, the gas stream, is composed of methane and carbon dioxide in about equal volumes. In order to provide high quality fuel gas, it is necessary to scrub the gas and remove from it the carbon dioxide. It is also necessary to dry the gas for distribution. The second stream, the slurry stream, is composed of an aqueous suspension of undigested organic matter which must be disposed of appropriately.

Bioconversion of Solid Waste and Sewage Sludge

Gas Treatment and Handling: The methane produced from the digester contains carbon dioxide, and perhaps some minor or trace quantities of hydrogen sulfide. These two acid gases must both be removed before the methane can be sold. A very common method of removing acid gases from natural gas is by contact with monoethanolamine (MEA). The sour gas is contacted with lean MEA in an absorber. The gas leaving this absorber has essentially no residual carbon dioxide or hydrogen sulfide. The MEA, which has been used to absorb the carbon dioxide and hydrogen sulfide, is drawn off the bottom of the absorber and sent to a stripper. Here sufficient heat is supplied to remove the acid gases and to regenerate the MEA as lean MEA. A portion (about 2% of the MEA circulating) is sent to an MEA reclaimer to maintain the purity of the absorbing solution.

Fresh MEA is added as makeup to replace that vaporized or chemically degraded in the process. Before the methane can be sold, it must also be dried. That is accomplished by means of a glycol dehydration process where the moisture is absorbed in dry glycol which is then regenerated and recirculated.

Effluent Handling and Disposal: During steady state operation of the waste digestion system, when there is no accumulation or depletion of nonreacting components, it is necessary to bleed-off the system continuously. Water and nondigestible matter must be removed from the system at a rate equal to their rate of feed. Because the digesters are reasonably well-stirred, the composition of the effluent is essentially the same as that of the contents. In addition to water and nondigestible matter, the effluent from the digesters contains undigested digestible solids, organic matter in solution, and a viable biological mass. Because the rates of digestion and gas generation are directly proportional to the concentration of microbial mass, it is advantageous to reclaim and recycle the biological mass back into the digesters. It is then possible to digest solids and produce gas within relatively compact digester volumes.

Such a process is only possible, however, if the viable biological mass can be separated from the nondigestible matter. If such a separation is possible, the viable mass is returned, in aqueous suspension, to the digesters while the remaining solids in suspension (as a slurry) are disposed of. The problem of disposal of the effluent remaining after the microbiological mass is recycled still remains. That effluent consists of a slurry of aqueous undigested matter. If no separation of microbiological organisms is carried out, the slurry containing undigested matter and organisms must be disposed of.

The disposal of final slurry is best carried out by separating the solids from the liquid and returning the liquid to the sewage treatment plant for subsequent treatment and final disposal. The solids, in the form of a moist sludge or cake obtained from a settling tank, filter or centrifuge, are then incinerated or possibly sent to landfill.

Computer Model Results

A computer program was formulated to account for the broadest possible scope of relevant factors, including capital costs based on engineering selection of equipment, credits and penalties associated with waste consumption and residue disposal, realistic financing and operating charges, and the most complete description of operating characteristics available. Representative values of operating and cost parameters were selected and utilized to establish a base-line process description. Confidence in the values selected was attained by conferring with recognized researchers, engineers, equipment manufacturers, and engineering design firms to verify and document the information and selections. The cost of gas produced by the base-line process was then calculated. A sensitivity study was conducted in order to determine the effect of variations from base-line conditions on gas cost. The effect on gas cost of plant site location and of the financing options available in government or private ownership was investigated.

Through this computer-aided optimization of the calculated cost of such a plant it was found that methane can be produced in such a process at a base-line cost of $2.09 per thousand cubic feet. This cost is economically acceptable when compared with the cost of natural or synthetic gas anticipated to develop in a short time. Operating energy requirements of the process consume the equivalent of only 37.5% of the gas produced, and the community obtains a storable fuel of versatile applicability.

A more detailed description of the computer model is given in the NTIS report, PB 238 068. For a complete bibliographic entry of the report, see p 222.

PFEFFER-DYNATECH ANAEROBIC DIGESTION SYSTEM

As a result of the work of Pfeffer at the University of Illinois and the pilot plant data obtained by the Dynatech Corporation, the Pfeffer-Dynatech Anaerobic Digestion System was proposed.

The process can be described best by use of Figure 3.1. This figure shows, by means of a block diagram, the major components of the Pfeffer-Dynatech anaerobic system. The waste handling consists of taking the solid waste as delivered to the plant, and shredding it by means of dry primary shredders. This reduces the maximum particle size to a range from 7.6 to 15 centimeters. A magnetic separator removes ferrous metals followed by a trommel screen to remove the fine grit.

The waste stream can then be passed through a hydropulper and then cyclone separators to reduce the size and remove nonferrous metals and glass. The second possibility is an air classifier for glass and nonferrous metal removal followed by a second dry shredder. The final ideal particle size is less than 2.5 centimeters. The remaining fraction of the stream, the organic matter, is then slurried with aqueous nutrients, pH control additives in the form of lime, and hydrogen sulfide control additives in the form of ferrous salts. This mixture is then fed into the digester.

The proposed system consists of a number of separate large digester tanks. Each tank is maintained at constant pressure by a floating cover. The circular tanks are constantly stirred to maintain an approximately constant density of suspension and thus uniform digestion. The digestion process is carried out in two steps. First the solids are solubilized by enzymatic action. Second, these soluble products are digested by the microorganisms to produce CH_4, CO_2, and small amounts of other gases such as H_2S. The cleanup of the gas to remove the acid fraction would utilize natural gas technology. The effluent production by the process, especially if it is continuous, produces a problem. Even though the biological mass could be reclaimed and recirculated into the digester, large amounts of

Bioconversion of Solid Waste and Sewage Sludge 81

FIGURE 3.1: BLOCK DIAGRAM OF THE PFEFFER-DYNATECH ANAEROBIC DIGESTION SYSTEM

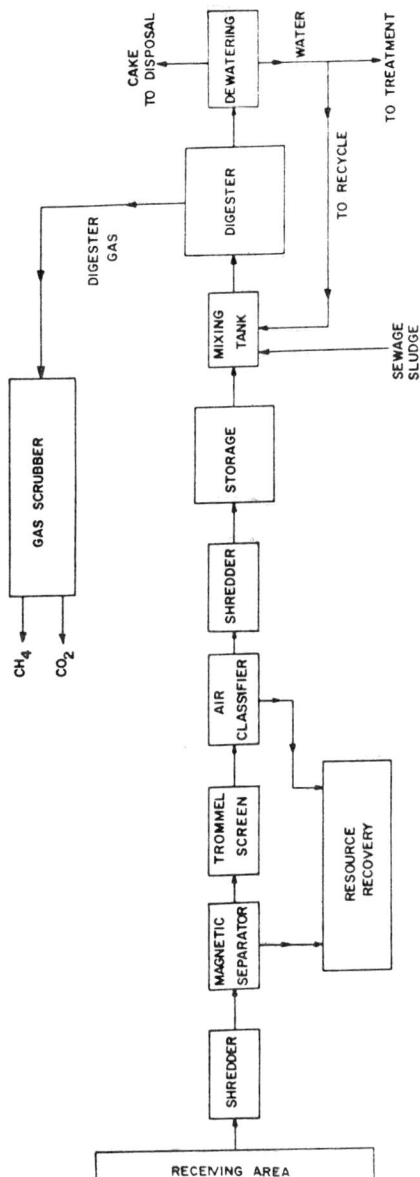

Source: N75-25292

water must be cleaned up and disposed of continuously. The other problem is to dispose of the remaining digested solids. The economics of the Pfeffer-Dynatech system are based on the experimental work of Pfeffer and the scaled up pilot plant data by the Dynatech Corporation and are given in 1974 dollars. The process costs and the resource recovery data are itemized in Table 3.5. The pilot plant information has been scaled up to 907 metric tons/day (1,000 tons/day). The following pertinent details are condensed from reference (1). The estimates will vary with conditions at a particular time and place. However, a base-line model for operating costs is developed. This model uses the following unit cost estimates:

Power	$2.78/$10^9$ joules	($0.01/kwh)
Steam	$0.95/$10^9$ joules	($1.00/million Btu)
Cooling water	$5.30/$10^6$ liters	($0.02/thousand gallons)
Lime	$27.56/metric ton	($25.00/ton)
Labor (28 men)	$5.63/man-hour	
Supervision (4 men)	$6.63/man-hour	
Disposal costs:		
Incineration of dewatered cake	$33/metric ton	($30/ton dry solids)
Landfill of inorganics	$1.38/metric ton	($1.25/ton)
Sewage treatment of water	$5.30/$10^6$ liters	($0.02/thousand gallons)

An estimate is made of the selling price of the methane to achieve a 15% return on the equity for a private investor. A standard gas utility method is employed assuming:

Period of depreciation	20 years
Depreciation method	5% on total capital-straight line
Federal income tax rate	48%
Percent interest on debt	9%
Debt/equity ratio	75%/25%

The cost model allows recovery credits only for ferrous metals as the inorganics are assumed landfilled. The other credits are for disposal of municipal sewage sludge and the refuse itself. Credits are as follows:

Ferrous metal	$18.74/metric ton ($17.00/ton)
Disposal of sewage sludge	$55.13/metric ton ($50.00/ton)
Disposal of municipal wastes	$11.74/metric ton ($10.65/ton)

The calculation of the methane selling price is as follows:

	10^9 Joules of Methane	10^6 Btu of Methane
Contribution of capital costs	$1.52	($1.602)
Contribution of operation costs	$1.49	($1.573)
Penalties:		
Filter cake disposal	$1.53	($1.610)
Wastewater treatment	$0.01	($0.010)
Inorganic waste disposal	$0.09	($0.100)
Total costs	$4.64	($4.895)

(continued)

	10^9 Joules of Methane	10^6 Btu of Methane
Credits:		
Scrap iron	$0.28	($0.300)
Sewage sludge disposal	$0.51	($0.540)
Municipal waste disposal	$2.91	($3.070)
Total credits	$3.70	($3.910)

The selling price of the methane is then the difference between costs and credits or $0.94/$10^9$ joules ($0.985/$10^6$ Btu). Any variation in the cost or credit items would change the selling price. The report indicates that the selling price of methane needed is most sensitive to the credit received for solid waste disposal. In reality, this credit will also vary widely from place to place. A wide spectrum of waste disposal costs is experienced across the country from $2.21/metric ton ($2.00/ton) in a landfill to as high as $28.66/metric ton ($26.00/ton) in an incinerator. To help a local region decide on the feasibility of this system for waste disposal, graphs such as Figure 3.2 might be developed.

FIGURE 3.2: METHANE PRICE VARIATION WITH LOCATION

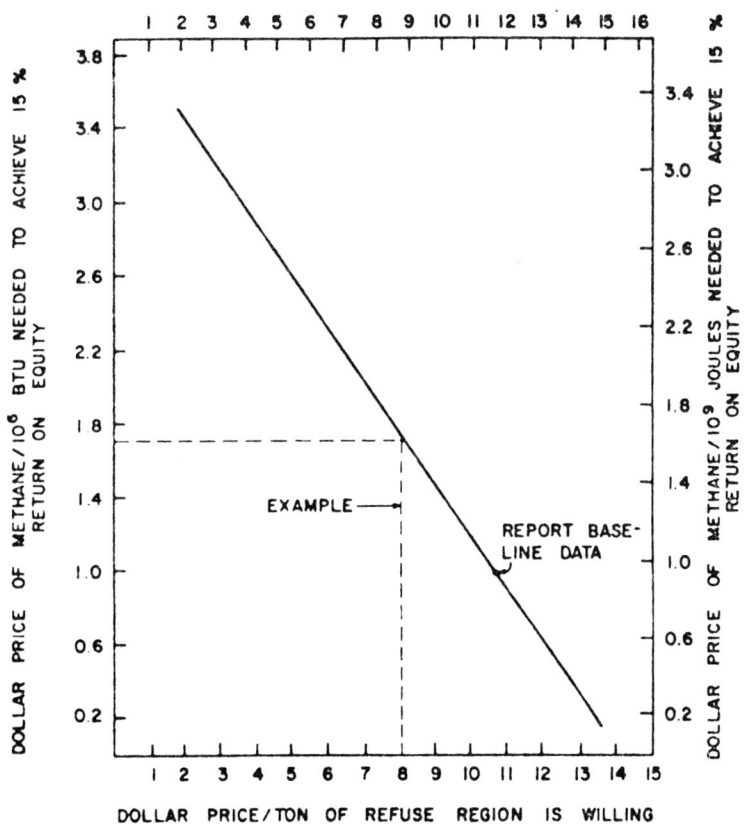

Source: N75-25292

TABLE 3.5: PFEFFER-DYNATECH METHANE GAS SYSTEM—PROCESS COST SHEET*

	DOLLARS	COMMENTS
CAPITAL COSTS (TOT. $)		
Land	(a)	(a) Not included
Preprocessing Eqmt	1,879,300	
Processing Eqmt	2,620,693	
Postprocessing Eqmt	3,062,218	
Utilities		
Building & Roads	500,000 (b)	(b) Building only
Site Preparation		
Engr. & R & D	502,947	
Plant Startup	75,403	(c) Contingencies
Working Capital	48,066	(d) Interest during construction
Misc.:	2,342,058 (c)	
	2,191,688 (d)	(e) Expense & profit - contractor
	1,760,316 (e)	
	320,246 (f)	(f) Plant equipment
TOTAL		
TOTAL	15,302,935	
OPERATING COSTS ($ PER YR)		
Maint. Material	194,817	
Maint. Labor	187,008	
Dir. Labor	249,344	
Dir. Materials	82,815	
Overhead	319,264	
Utilities	356,818	
Taxes (Local)		
Insurance	350,670	
Interest		
Disposal of Residue	1,969,424	
Payroll Benefits		
Fuel		
Misc.:		
TOTAL	3,770,160	
CREDITS ASSUMED ($ PER YR)	5,714,791	

	DOLLARS/YR.	COMMENT
Fuel:		(a) Not calculated in report-calculation done based on 330 day/yr and $.935/10^9 joules ($.987/10^6 Btu-report figure)
Liquid		
Gas Methane	1,240,000 (a)	
Solid		
Power:		
Steam		
Electricity		
Hot Water		
Magnetic Metals	343,195 (b)	(b) 18.70/metric ton (17.00/ton)
Nonmagnetic Metals	(c)	(c) Glass & non ferrous metals assumed landfilling
Glass	(c)	
Ash		
Paper	(d)	(d) Light organics landfilling
Other: Sewage sludge	618,502 (e)	(e) $55.00/metric ton (50/ton)
Disposal credit	3,513,094 (f)	
TOTAL ($ PER YR.)	5,714,791	(f) City pays 11.70/metric ton (10.65/ton)

*Reference (1)

Source: N75-25292

The needed selling price of methane is shown vs the price per ton a municipality is willing to pay for refuse disposal. Based on this example graph, a community willing to pay $8.82/metric ton ($8.00/ton) for refuse disposal would necessitate the private investor selling his methane at $1.61/$10^9$ joules ($1.70/$10^6$ Btu). If the current or near term price of natural gas in this area is only $0.95/$10^9$ joules ($1.00/$10^6$ Btu), the feasibility of this disposal system is clearly questionable.

Since carbon dioxide is also produced, the credit situation could be improved if a suitable market were available. The plant theoretically produces 227 metric tons per day (250 tons/day) of carbon dioxide and 86 metric tons/day (95 tons per day) of methane. Chemical companies are currently paying about $8.821 per metric ton ($8.00/ton) for 98% carbon dioxide with no sulfur content. If the carbon dioxide is then purified to 99.9%, it may be sold for up to $149 per metric ton ($135/ton). This added credit, depending on the purity of carbon dioxide, would shift the graph down to the left indicating a lowered selling price for methane.

Another set of credits may be obtained by adding other resource recovery equipment for the recovery of aluminum and glass. A lower cost structure would also be obtained if public financing is used. The energy needs for the system are about 35% of the methane produced. A tradeoff analysis must be made concerning use of the methane or purchase of power. It is noted that the costs in this report are for a hypothetical 907 metric ton/day (1,000 ton/day) system.

SUPPORTING WORK FOR DYNATECH

Support to the Dynatech program has been furnished by the Massachusetts Institute of Technology and the University of Massachusetts. Both of these academic programs were initiated as experimental projects to provide technical support to the operation by Dynatech of the originally planned pilot plant for the anerobic digestion of solid waste. The studies at the University of Massachusetts were to include product gas scrubbing evaluation, kinetics studies, development of microorganism separation techniques, digester operation, and investigation of methods for monitoring microorganism activity. The study at the Massachusettts Institute of Technology was to investigate thermophilic microorganism development.

University of Massachusetts

Five supporting studies were conducted at the University of Massachusetts. Four 22-liter anaerobic digesters were operated over a period of six months. An examination of the use of tracer studies to determine rate constants was undertaken. Initial investigations were conducted to develop an important tool to monitor digester performance. Alternate methods of concentrating microorganisms were examined. Finally, an investigation was conducted to evaluate methods of gas processing to produce a pipeline quality product. Feed to the digesters was 4.6% total solids. Solids retention time was 20 days. Digester performance was poor, and long-term stability was difficult to achieve due to equipment failures and inadequate digester control. The digesters were plagued by wide swings in pH and severe build-up of deleterious ions. Radioactive tracer studies provide a valuable tool for use in determining kinetic rate constants characterizing biological processes. A mathematical analysis of the relationship between the rate of disappearance of the tracer and the first-order reaction rate constant used on the Monod equation was carried out (2). As a result of this analysis, it may be shown that tracer studies yield a pseudo first-order reaction constant which is concentration dependent. Therefore multiple experiments using digesters operating at steady state but at different concentrations and retention times are necessary in order to properly evaluate each of the coefficients in the Monod equation.

Adenosine triphosphate (ATP) is a characteristic component of living cells. The concentration of ATP is a function of the growth conditions of the cell and may be correlated with biomass and microbial activity. The research at the University of Massachusetts was aimed at developing an on-line instrument to measure ATP concentration in an anaerobic digester, providing a more responsive instrument with which to evaluate digester performance. Levels of ATP were determined by measuring a bioluminescent reaction using prepared extracts of firefly tails as a source of lucoferin-luciferose. The emitted light was measured using a photometer and recorded on a chart recorder. The method proved to be a reliable indicator of digester methane production, with increases and decreases of methane production being preceded by similar changes in ATP concentration.

Three sets of experiments were conducted to evaluate the feasibility of various microorganism separation techniques. The objective of microorganism separation is to selectively separate the microorganisms from the digester effluent and then to concentrate and recycle them to the digester, increasing the microbial population. Initial experiments with a dispersed gas flotation column were inconclusive. The data collected were erratic and poor digester performance resulted in large quantities of undigested fibrous material which tended to mask results.

Accordingly, a second set of tests were run using a colony of pure *E. coli* bacteria and two nonionic surfactants, Spanzo and Atlox 1045. Again the data were inconclusive, but neither surfactant appeared to concentrate the microorganisms in the foamate (2). A third set of experiments was conducted to evaluate the potential for elutriation as a microorganism separation technique. A strainer of 25 mesh steel was used. By population counts it was determined that approximately 70% of the microbial population was washed through on the first wash. This could be concentrated using centrifugation or other techniques and recycled to the digester to increase the microbial population.

Additional studies were undertaken to determine a gas processing technique suitable for use with the digester off-gas to produce a pipeline quality product. The process selected, absorption of the CO_2 and H_2S by monoethanolamine (MEA) followed by moisture removal in a glycol dehydration unit, was designed to meet the operating requirements of the 1,000 tpd facility and incorporated into the economic model.

Massachusetts Institute of Technology

Part of the Dynatech program was to examine innovative methods of digestion in an attempt to lessen the biological and engineering problems which have restricted the broad and large scale use of this process for fuel gas production. One such approach was to employ thermophilic culture conditions as a means of improving some aspects of the process. It is in this area, thermophilic anaerobic conversion of solid waste to fuel gas, that several aspects of the problems have been explored. Thermophilic digestion offers several advantages over traditional mesophilic digestion. These are:

1. Increased rate of digestion
2. Decreased fluid viscosity
3. Decreased biomass formation
4. Increased conversion of waste to gas
5. Absence of pathogen formation

The primary drawback to the approach is the need to supply heat to maintain the temperature above 50°C. Capitalizing, however, on the positive attributes, there is interest in attempting to minimize the retention time in a thermophilic digester, while at the same time increasing the solids loading and in this way attempting to maximize productivity. The work undertaken at MIT had three specific objectives.

1. Further develop and characterize a thermophilic anaerobic population of bacteria able to convert solid waste to methane gas.
2. Determine the effect of residence time, recycle and temperature on the extent of solid waste degradation, the biological mass formed and the quantity and quality of gas produced.
3. Examine the possibility of treating (e.g., by chemical hydrolysis) the cells formed in the digester and then recycling hydrolyzed cells to promote growth in the digester.

Three 5-liter fermenters were operated during a 12-month period on a mixture of sewage sludge and shredded newsprint at thermophilic temperatures of 55°C. Each fermenter was continuously agitated by four-bladed impellers and each tank was baffled to promote mixing. Daily feeding and sampling was accomplished through a one-inch opening in the top of each vessel. Samples were obtained by rapidly withdrawing material by suction.

A self-sustaining microbial colony was achieved within two months using a continuous enrichment process. The feedstock contained 5% total solids with a 9:1 ratio of shredded newsprint to sewage solids. Retention time was 24 days. After an initial acclimation period, experiments were conducted with an 8% solids feed at retention times of 5, 10, 15 and 24 days. Additional experiments were conducted at 3 day retention time, 5% total solids feed concentration.

The results from the experiments conducted lead to several conclusions. A stable microbial colony can be established at thermophilic temperatures with relative ease on one of the principal components of solid waste. Operating digesters were developed within three retention periods. The populations were observed to be reasonably stable during the experimental program and recovered well from those cases where operational problems were encountered. On the other hand, digester performance was considerably poorer than expected. Apparent volatile solids reductions for 5, 10 and 15 day retention times were approximately 50%, as computed on the basis of an input/output analysis of the digester slurries. However, potential sampling errors make this type of analysis unreliable. For example, small differences in the volatile determination of the solids will result in wide variation in the apparent volatile solids reduction. An alternative procedure is to compute the volatile solids reduction on the basis of the gas output quantity and composition. Using this procedure, the apparent volatile solids reduction is approximately 7 to 8%. The discrepancy between the two results may be explained in several manners. Sampling errors can play a large factor in both analyses. The input/output analysis requires one to obtain a representative sample from a completely mixed digester. Precise volatile determinations are also required, as noted above. The gas production analysis requires a highly accurate, leak-free gas monitoring system.

Proper experimental techniques are particularly important when dealing with such small reactors (5 liters). The substrate chosen, newsprint, may be expected to be more difficult to digest than other organic wastes. Newsprint typically contains 40 to 50% lignins which are volatile but not readily biodegradable. This would lower the volatile solids reduction. Investigations at MIT have shown another reason for poorer than expected performance. Newsprint often contains biocides to reduce degradation. The suppliers confirmed the presence of biocides in the newsprint used in the experiments.

In addition, it appears that the ink used may be toxic to the microorganisms. Experiments run with deinked newsprint showed substantially improved digester performance. It should be noted that such toxicity problems have not been noted in similar experiments conducted at the University of Illinois. However, the water separation technique employed to separate inerts from the organic materials (3) may help to wash the toxic materials from the feedstream. Further experiments would be desirable to assess the need for such processing in large-scale facilities.

Although the digester performance was not as good as expected, several significant conclusions may be drawn from the experimental program. The minimum solids retention time for the thermophilic culture was confirmed to be 4 to 5 days, as expected from previous experimental work. An analysis of the volatile acids in the digesters showed that the relative concentrations of propionic, acetic, and butyric acids were 67, 23, and 10% respectively, regardless of retention time.

This observation suggests that there is no change in the composition of the microbial colony as a function of retention time, at least within the range examined. Finally, an increase in the methane composition of the gas produced was noted with decreasing retention time. It has been shown that this variation can be attributed to more of the CO_2 produced in the reaction being carried out in the effluent stream as dissolved CO_2. If the methane concentration in the digester gas can be raised, capital and energy savings will result from the reduction in gas scrubber capacity. However, this effect is not expected to be as great as the experimental data suggests if higher overall gas production rates are achieved.

OTHER DIGESTION STUDIES

University of Pennsylvania

At the Bioconversion Energy Research Conference of June 1973, M. Wolf and R. Keenan described work being done at the Towne School of Engineering at the University of Pennsylvania. According to them, the priorities that have been established to accomplish the goal of economic anaerobic methane production are studies of fuel gas generation, system operation and raw material generation and the greatest emphasis has involved the digestion process itself. The problems involved with digester operation, essentially low loadings and instability, result in higher costs. It is necessary to thoroughly understand digester dynamics in order to overcome these problems and to optimize methane productivity. Research concerning digester technology has centered on, and will continue to emphasize, conventional anaerobic digestion, the potential of two-stage digestion and a basic understanding of digester microbiological population dynamics.

Conventional Digestion: Conventional digesters were operated at the University of Pennsylvania for many months. Experiments have been conducted utilizing mesophilic (37°C) and thermophilic (48°C) cultures. Both types have operated under identical conditions [0.31 pound volatile solids (VS) per cubic foot per day, loading; 10-day detention time; 5% slurry of ground dog food as feed material]. Results indicate that gas production is greater in the mesophilic digesters (9.5 cubic feet of gas per pound of dry organic matter) than in the thermophilic digesters (8.6 cubic feet of gas per pound). In both cases, methane comprises about 60% of the total gas produced. The significance of this is that operating costs are lower for the mesophilic digester since a reduced temperature is maintained. Additional research was carried out to investigate gas production at higher loadings.

Two-Stage Digestion: An experimental program was conducted to evaluate the potential of separating the acidogenic bacteria from the methane producers as a means of improving digester performance. It was anticipated that successful separation will provide greater stability, and the possibility of decreased detention time and, hence, costs.

A Stage 1 digester was initiated by overloading a conventional thermophilic digester and by maintaining pH 6 with sodium hydroxide or calcium hydroxide. The digester was operated for six weeks, and at a loading of 0.52 pound VS per cubic foot per day for the last four weeks. Gas production stabilized at 1 cubic foot per pound VS, or approximately 11% of the gas generation observed in the conventional digesters. The composition of the gas was about 98% carbon dioxide. Volatile acids concentration was steady at 10,000 parts per million (as acetic acid). The volatile acids present included formic, acetic, propionic, isobutyric, n-butyric, isovaleric, n-valeric and a higher acid, most probably caproic.

Two attempts have been made to establish a Stage 2 culture by feeding Stage 1 effluent to a conventional thermophilic culture. The first attempt failed, with gas production falling rapidly. After reestablishing the mixed (Stage 1 and Stage 2) culture, the procedure was repeated. The Stage 2 digester received Stage 1 effluent for about 10 days. At the loading conditions of 500 to 700 milliliters of Stage 1 effluent per day, gas production in the Stage 2 tank averaged 10 cubic feet of gas per pound VS added to the Stage 1 tank. This represents an increased gas production relative to the conventional digesters. The gas generated consists of 70 to 75% methane. Consequently, methane productivity (7.4 to 7.7 cubic feet of methane per pound VS) is greater than in the conventional digester (5.3 to 5.7 cubic feet of methane per pound VS). Gas purification costs would, as a result, be reduced since there is less carbon dioxide to remove. These results indicate that two-stage digestion is achievable, and that methane productivity and the Btu content of the fuel gas are increased by the use of two-stage digestion.

University of California

At the same conference, C.G. Golucke described work done at the Sanitary Engineering Research Laboratory at the University of California (Berkeley). The approach was to determine the digestibility of the major components of domestic refuse, as well as the digestion potential of the mixed wastes as they come from the collection (garbage) truck after removing metal and glass.

Experiments involved with determining the digestibility of the individual components of refuse were done on a laboratory scale. That is, the digesters used in this phase of the study had only a one-gallon capacity per digester. They were equipped for stirring both by mechanical means and by recirculation of the digester gas through the culture. Temperature was maintained at 37°C. Detention periods of 15 and 30 days were applied. The pilot plant digester had a capacity of 400 gallons. Mixing in the pilot digester was done by recirculating its contents. It, too, was maintained at 37°C.

Components digested individually were Kraft paper (pulp), newsprint (pulp), green garbage, grass clippings, and sawdust (white fir and Monterey pine). Nitrogen was added in the form of raw sewage sludge or chicken manure to bring the C:N ratio down to a suitable level. The studies included an analysis of the kinetics of the hydrolysis of cellulose in anaerobic digestion.

Back in the 1930s when the home garbage grinder was first introduced, a great deal of research was carried on with the objective of determining effects, adverse or beneficial, on the culture in the sewage treatment plant digester when ground garbage was introduced into the culture. A conclusion reached by the majority of the early researchers was that a definite upper permissible limit existed with respect to the fraction of the total culture in the form of garbage. At high levels, according to them, acid formation would be excessive, and the digester would fail.

It has been found that after a period of acclimation, green garbage could serve as the sole feed to a digester without any adverse effect. Once acclimated, an average of 65% of the incoming volatile solids was destroyed, with an accompanying gas production of 12 to 16 cubic feet of gas per pound of volatile solids introduced. Methane constituted from 50 to 55% of the gas.

In experiments with paper pulp added to raw sewage solids, it was found that degradation of the cellulose content was as much as 90%, and that loadings of paper up to 60% of the total solids in the digester were feasible, i.e., until the resulting C:N ratio was 45:1. Digester failure occurred when the C:N ratio reached 52:1. Experiments with paper pulp to which chicken manure was added as a source of nitrogen involved a C:N ratio as high as 70:1. Despite the high ratio, digestion proceeded without difficulty, and approximately 90% of the cellulose was destroyed.

In the experiments on the digestion of newspaper cellulose, with the detention period of 30 days, 83% of the total cellulose in a mixture of 10% newspaper and

90% sewage sludge was destroyed. About 55% of the cellulose in the newspaper portion of the mixture was digested. With the proportion of newspaper stepped up to 20%, 73% of the total cellulose was destroyed, and 50% of the newspaper cellulose. Increasing the newspaper content of the mixture to 30% resulted in a 63% destruction of the total cellulose, and 42% of the newspaper celllulose. The composition of the gas from a digester receiving only raw sewage sludge was almost identical with that of gas produced by digesters receiving newspaper cellulose.

Grass clippings proved to be 73% digestible. The average destruction of the cellulose in the grass clippings amounted to 79%. Experiments with wood showed that the addition of wood to a digester would exert neither a chemically nor a physically adverse effect on the culture. Wood underwent very little decomposition in the digesters. In essence it acted as an inert material. Perhaps the imposition of detention periods longer than those applied in the present study might lead to the eventual decomposition of the wood. In experiments involving the digestion of a synthetic refuse (components of the same nature and in the same concentrations as found in municipal refuse), the total solids reduction averaged 44%; and volatile solids reduction, 52%. The calculated amount of destruction of the cellulose in the synthetic refuse was 55%. Gas production was about 56% that of a digester receiving raw sewage sludge. No adverse effects were noted with the C:N ratio adjusted to a high of 48:1.

Hydrolysis of Cellulose: Degradation of the cellulose in the anaerobic digester process was explored in an investigation concerned with the kinetics of cellulose degradation under anaerobic conditions. The report of the investigation is summarized as follows:

The specific objectives of the investigation were:

1. To determine the hydrolysis rate of cellulose and the corresponding microbiological kinetic constants in continuous-flow anaerobic fermentors.
2. To evaluate the applicability of the Michaelis-Menten kinetic model to a particulate substrate, cellulose, in a continuous-flow reactor system.
3. To determine whether or not the hydrolysis rate of cellulose in anaerobic fermentation processes is growth associated, that is, if there is a distinct relationship between the rate of hydrolysis and the rate of growth of the mixed culture system.

Specific conclusions drawn from the investigation were:

1. The results of the growth and kinetic study indicated that the hydrolysis rate of cellulose in anaerobic fermentation can be accurately characterized by the Michaelis-Menten kinetic model and the cell continuity equation as follows:

$$q = \frac{q_{max} S_1}{K_s + S_1} = \frac{S_0 - S_1}{\theta X_1} \quad \text{(Michaelis-Menten Equation)}$$

$$\frac{1}{\theta} = Yq - k_d \quad \text{(Cell Continuity Equation)}$$

2. The kinetic constants and coefficients observed can be summarized as follows:

X_1 Cell Conc.	Y Yield Constant mg X_1 Produced per mg Cellulose Hydrolyzed	k_d Decay Rate day^{-1}	K_s M-M Constant mg Cellulose per liter	q_{max} Maximum Specific Growth Rate day^{-1}	q Cell Generation Time days
Organic Nitrogen Conc. as X_1 mg Org-N/l	0.015	0.01	4,422	0.360	1.93
Volatile Suspended Solids Conc. as X_1 mg VSS/l	0.104	0.005	4,480	0.334	2.07
Dehydrogenase Activity as X_1 mg TF formed hr^{-1}/l	0.007	0.047	523	0.160	4.33

3. The maximum hydrolysis rate of cellulose in a continuous-flow, noncellular recycling fermentor was found to be:

   q_{max} = 23.4   mg cellulose hydrolzyed/mg Org-N-day

   = 3.21 mg cellulose hydrolyzed/mg VSS-day

   = 22.9 mg cellulose hydrolyzed/mg TF formed hr^{-1}-day

4. The cell washout time or the minimum mean residence time of the mixed culture studied was about 3 days; and the calculated generation time observed, about 2 days (based upon both organic nitrogen and VSS concentrations to represent the cell concentration X_1).

5. When the residence times of the system were decreased, the total volatile acids, as well as the carbon concentration in the filtrate of the system, increased only slightly. These conditions, together with the results observed in cellobiase activity measurement, indicate that the cellulose hydrolysis to cellobiose not only is the limiting step of the cellulose hydrolysis process, but it also is the rate-limiting step in overall cellulose fermentations.

6. Based on the hydrolysis rates observed in this study, the calculated data indicate that cellulose hydrolysis is not the rate-limiting factor in the domestic sewage sludge digestion process.

7. The relationships of the net growth rate $1/\theta$ and q, in terms of cellulose hydrolysis rate, or in terms of the gas production rate, follow the same general pattern over the range of $1/\theta$ studied. Therefore, it appears that both the hydrolysis rate and the methane fermentation rate are growth associated.

Bioconversion of Solid Waste and Sewage Sludge

> 8. Dehydrogenase activity appears to be a suitable parameter to measure in operating and controlling anaerobic fermentation systems. It appears to be one of the most sensitive indicators of digestion failure.

Pilot Plant Experiments: The major difficulty in operating the pilot unit proved to be that the mixing system in the digester was not adequate to prevent the formation of a thick scum layer. With the large amount of floatable material in refuse, it is likely that an extensive mechanical mixing system rather than gas mixing or an effluent recirculation system would be needed to avoid this problem. The scum layer did not interfere with the overall functioning of the digester, except that it excluded a significant quantity of raw refuse from contact with the bacterial population. In general the pilot digester performed to the degree of efficiency predictable from laboratory studies. The results may be summarized as follows:

> 1. With refuse of the type used in the study, mixtures of refuse and sewage sludge may be digested without difficulty in proportions approaching 60% refuse.
> 2. The C:N ratio proved to be the limiting factor with respect to the size of the refuse fraction.
> 3. The coarsely shredded refuse used in the experiments were digested with reasonable efficiency, despite the well-known fact that efficiency of decomposition increases with decrease in particle size.
> 4. Destruction of volatile solids in the influent total solids amounted to 66.8%. Removal of volatile solids was 82.8%.
> 5. The difference between the amount destroyed and the amount removed may be attributed for the most part to diversion of material to the scum layer.
> 6. Cellulose destruction amounted to nearly 80% and cellulose removal to 94%.
> 7. Removal of total solids amounted to 74.2%. However, due to the high proportion of inorganic matter in the refuse, total solids destruction was only 39.8%.
> 8. Gas production averaged 10.0 cubic feet per pound of volatile matter destroyed, and 7.0 cubic feet per pound introduced into the digester.
> 9. The methane content of the gas increased from 55% at the start of the run to about 60% in the final month.

Clemson University

Another contributor at the conference was J. Andrews who described work done at the Department of Civil Engineering of Clemson University on dynamic modeling and control strategies for the anaerobic digestion process. He said that the anaerobic digestion process has in general not enjoyed a good reputation because of its poor record with respect to process stability as indicated through the years by the many reports of sour or failing digesters. The major problems with the process appear to lie in the area of process operation as evidenced by its more successful performance in large cities where skilled operation is more prevalent. There is a great need for control strategies, based on dynamic models, to put process operation on a more quantitative basis. The implementation of such strategies should result in a decrease in the frequency of process failure and permit the optimization of process performance. Dynamic models would also be of value in improving process design since they would allow comparison of the different versions of the process with respect to stability. The incorporation of modern control systems would also improve process stability and decrease the need for oversizing. The dynamic model proposed for the process was developed from material balances on the biological, liquid, and gas phases of a continuous flow, complete mixing reactor. The components on which material balances are made are given below:

----------------------Phases----------------------

Biological	Liquid	Gas
Organisms	Volatile acids	Carbon dioxide
	Conservative toxic agent	Methane
	Cations	
	Bicarbonate	
	Dissolved carbon dioxide	
	Methane	

There are strong interactions between the phases as well as within each phase. These interactions must be considered if the model is to predict the dynamic response of the five variables most commonly used for process performance: (1) volatile acids concentration; (2) alkalinity; (3) pH; (4) gas flow rate; and (5) gas composition. The following relationships were used to express these interactions on a quantitative basis:

1. Yield coefficients
    a. Mols organisms produced per mol volatile acid utilized
    b. Mols carbon dioxide produced per mol organisms produced
    c. Mols methane produced per mol organisms produced.
2. Kinetics of organism death due to a conservative toxic agent.
3. Inhibition function for relationship between organism growth rate and unionized acid concentration.
4. Equilibrium relationship between ionized acid, unionized acid, and pH.
5. Equilibrium relationship between dissolved carbon dioxide, bicarbonate, and pH.
6. Charge balance on ionic species in solution.
7. Henry's Law.
8. Mass transfer equation for transfer of carbon dioxide across the gas-liquid interface.

The model was kept as simple as possible by considering the conversion of volatile acids to methane and carbon dioxide as the rate limiting step. It is also assumed that there is no lag phase, endogenous respiration, or inhibition by-products. The model is restricted to a pH range of 6 to 8 and does not consider the precipitation or dissolution of solid chemical phases such as calcium carbonate. Two key features of the model are the use of an inhibition function in lieu of the Monod function to relate volatile acids concentration and specific growth rate for the methane bacteria and consideration of the unionized fraction of the volatile acids as both the growth limiting substrate and inhibiting agent. The use of an inhibition function is an important modification since it enables the model to predict process failure by high concentrations of volatile acids at residence times exceeding the washout residence time. Consideration of the unionized fraction of the volatile acids as the inhibiting agent resolves the conflict which has existed in the literature as to whether inhibition is caused by high volatile acids concentration or low pH. Since the concentration of unionized acids is a function of both total volatile acids concentration and pH, both are therefore of importance.

Digital computer simulation studies provide qualitative evidence for the validity of the model by predicting results which have been commonly observed in the field. Among the results predicted by the model are: (1) at steady state, an increase in the alkalinity concentration in the digester results in an increase in the operational levels of pH and volatile acids; (2) failure of the process can occur through hydraulic, organic, and toxic material overloading; (3) the course of failure, as evidenced by the behavior of the operational variables, pH, alkalinity (HCO_3^-), volatile acids concentration, and gas composition is qualitatively the same as that observed in the field; (4) stopping or reducing the flow to the reactor, the addition of base, or recycle of sludge from a second stage reactor are effective techniques for curing failing digesters.

Hybrid computer simulations were used to analyze process stability by simulating digester overloading and observing what changes in design and operational characteristics provided the best buffer against process failure. The analysis procedure involved making a change in a digester parameter, such as increasing the residence time (θ), followed by simulating larger and larger step increases in digester loading until failure occurred. By plotting the locus of points of critical substrate loading rates versus reactor residence time or other parameter, it was possible to obtain a semiquantitative measure of digester stability.

In addition to the increases in stability which are obtained by increases in residence time or alkalinity, stability also increases sharply with an increase in the concentration of methane bacteria in the digester. This increased concentration can be attained by increasing the influent substrate concentration (sludge thickening) or by recycle of concentrated sludge from a second stage. It is significant that three of the measures for improving stability, increased residence time, alkalinity, and influent substrate concentration, can be attained by sludge thickening. Other simulations indicated that process stability could be enhanced by the incorporation of suitable control systems.

A variety of control signals, controller modes, and control actions are available for use in developing control strategies for the anaerobic digester and were investigated using the hydrid computer. The simulations indicated that the most effective control strategy was directly dependent on the type of overloading. The recycle of gas from which carbon dioxide has been scrubbed, a control action proposed as a result of this research, and base addition, both using pH as the feedback signal, were best suited for the correction of organic overloading. A simple on-off controller mode was used to control the response to a step forcing in influent substrate concentration insufficient or sufficient to cause process failure. Failure by an overload of toxic materials was best prevented by the recycle of concentrated sludge from a second stage using the rate of methane production as a feedback signal. The control signal, rate of methane production, appears to be one of the best indicators of digester condition with respect to overloading with toxic materials. The dynamic model, process stability characteristics, and control strategies summarized herein are discussed in more detail in the publications of Andrews (4)(5).

Addition of Coal to Sewage Sludge to Increase Methane Production

In U.S. Patent 3,640,846 (February 8, 1972) assigned to the U.S. Secretary of the Interior, G.E. Johnson discloses that a mixture of sewage and coal substantially increases the production of methane by anaerobic digestion.

In the process, a sewage sludge containing methane-producing anaerobic bacteria is admixed with particulate coal in a vessel from which air is excluded. Since sewage solids decomposition for methane production is best attained at about 95° to 100°F, the vessel is preferably maintained at this temperature, although a wider temperature range (e.g., about 50° to 150°F) can be employed.

Any methane-producing anaerobic bacteria heretofore employed to produce methane from sewage can be employed. Exemplary bacteria suitable for the purposes are found in the genera Methanobacterium, Methanococcus and Methanosarcina.

Suitable coal-to-sludge weight ratios are about 1:100 to about 1:2. Generally, there is no limitation as to the maximum amount of coal that can be added. Although round-shaped tanks are usually employed to produce methane from sewage sludge, the operation is also readily carried out in rectangular or oblong vessels, either horizontally or vertically disposed, through which the coal and sludge are concurrently fed.

The lower ranking coals such as lignite and subbituminous A, B and C are preferably employed because they contain more cellulose-like material upon which the bacteria may feed. However, higher ranking coals such as low volatile bituminous (LVB) and high volatile A bituminous (HVAB) can be employed.

Whatever coal is employed, at the termination of gas production the residual coal can be employed as fuel. A coal particle size of about −60 mesh is suitable although finer particulate coal (e.g., −325 mesh) is preferred since more coal surface is exposed.

Example 1: Three 2,000 ml samples of activated sewage sludge which contained anaerobic methane-producing organisms were placed in separate glass flasks, from which air had previously been flushed from each flask system with an inert gas (helium or nitrogen).

To one of the flask was added an individual 100 gram sample of LVB coal (−325 mesh); to another, 100 grams of HVAB (−325 mesh) coal was added. The third flask contained no coal.

The contents of each flask were slowly stirred. The product gas produced during digestion (at 95° to 100°F) in each flask was periodically quantitatively and qualitatively analyzed. The results are shown on the following page.

TABLE 3.6: ADDITION OF LVB AND HVAB

Hours of Operation	Test Results with Sludge Only Gas Analysis (vol %)		Cumulative Gas Volume (l)	Test Results with LVB Coal Gas Analysis (vol %)		Cumulative Gas Volume (l)	Test Results with HVAB Coal Gas Analysis (vol %)		Cumulative Gas Volume (l)
	Methane	Carbon Dioxide		Methane	Carbon Dioxide		Methane	Carbon Dioxide	
300	-	-	-	82	18	3.9	-	-	-
400	77	23	2.4	84	16	7.3	72	28	2.5
600	-	-	-	87	13	12.8	83	17	6.5
650	82	18	4.5	83	17	13.5	78	22	7.2
800	-	-	-	85	14	14.6	80	20	11.0
1,100	79	21	5.7	-	-	-	-	-	-
1,300	82	18	7.7	87	13	15.7	83	17	12.6

Source: U.S. Patent 3,640,846

Example 2: Two 2,000 ml samples of activated sewage sludge which contained anaerobic methane-producing organisms were placed in separate glass flasks from which air had previously been flushed with an inert gas (helium or nitrogen). Initially, no coal was added to either flask. Lignite was later added to both flasks at the times shown below. The contents of each flask were slowly stirred. The product gas produced during digestion (at 95° to 100°F) in each flask was periodically quantitatively and qualitatively analyzed. The results were as follows:

TABLE 3.7: ADDITION OF LIGNITE

Hours of Operation	Test A Gas Analysis (volume percent)		Cumulative Gas Volume (liters)	Test B Gas Analysis (volume percent)		Cumulative Gas Volume (liters)
	Methane	CO_2		Methane	CO_2	
100	28	72	0.6*	-	-	0.8
200	44	56	1.3	88	12	1.6
250	52	48	2.2	90	10	2.6
325	70	30	3.2	95	5	3.5
400	74	26	4.0	95	5	4.1
500	74	26	5.2	95	5	4.5
750	78	22	6.3*	-	-	4.8
900	69	31	7.8	-	-	4.9**
1,000	56	44	8.4*	91	10	5.4
1,100	-	-	8.6	-	-	5.5

*200 grams lignite (–325 mesh) added in Test A at the end of 100 hours, 750 hours and 1,030 hours.
**Gas production in Test B ceased after 900 hours; 100 grams lignite (–325 mesh) added at the end of 900 hours which reinitiated gas production within minutes.

Source: U.S. Patent 3,640,846

Example 3: 3,000 ml of sewage sludge which contained anaerobic bacteria was placed in an air-free flask at 95° to 100°F. Product gas was periodically quantitatively and qualitatively analyzed. After 365 hours when gas production ceased, 250 grams of powdered lignite were added to the flask. The product gas was again qualitatively and quantitatively analyzed. The results are shown on the following page.

TABLE 3.8: REACTIVATION BY LIGNITE ADDITION

Digestion Time (hours)	Gas Analysis (vol %)		Cumulative Gas Volume (liters)
	Methane	Carbon Dioxide	
25	60	40	0.7
100	72	23	1.3
150	78	22	3.0
200	80	20	4.2
250	85	15	5.5
350*	68	33	6.0
500	55	45	6.5
700	67	33	7.1
800	76	24	7.8
1,000	95	5	8.4
1,440	91	9	8.9

*Gas production ceased after 365 hours, 250 grams of lignite (-325 mesh) was then added reinitiating gas production.

Source: U.S. Patent 3,640,846

As can be seen from Table 3.6, the addition of coal has a pronounced effect on total gas production. The LVB coal more than doubled gas production while the HVAB coal almost achieved the same result. Table 3.7 shows that lignite also has a similar effect. Tables 3.7 and 3.8 both show that the addition of coal at a point where gas production has ceased will substantially reactivate such production.

At sewage facilities internal combustion engines are generally used to drive auxiliary equipment such as sewage pumps or blowers which provide air for the aeration tanks, etc. Heretofore, the generation of sewage gas (CO_2 + CH_4) at such facilities has generally been insufficient to completely operate those engines on sewage gas, and the fuel has to be supplemented with natural gas. This process provides a method of generating sufficient gas at such facilities to completely fuel all these auxiliary engines. An additional benefit is that coal which has been previously used elsewhere in a sewage treatment system as, for example, an adsorbent, settling agent, filter aid, filtration agent or the like, can be reused in the digestion process to generate gas.

Synergistic Production of Methane from Refuse and Sludge

G.E. Johnson also disclosed in PB 220 821 that the combination of sewage sludge and refuse synergistically produces methane-containing gas by anaerobic digestion. Municipal sewage sludge means the sludge material removed from sewage in a municipal sewage primary treatment system, and municipal refuse means the garbage, trash, etc., collected from homes, businesses, institutions, etc., in a community, which material is commingled in the collection process, and usually dumped in designated refuse areas in such a commingled state.

Referring to Figure 3.3, influent sewage **1** enters primary sedimentation zone **2** to separate into supernatant liquor **3** and sludge **4**. The sludge usually consists of a suspension containing about 1 to 10 weight percent solids. Shredded or

FIGURE 3.3: SYNERGISTIC PRODUCTION OF METHANE FROM REFUSE AND SLUDGE

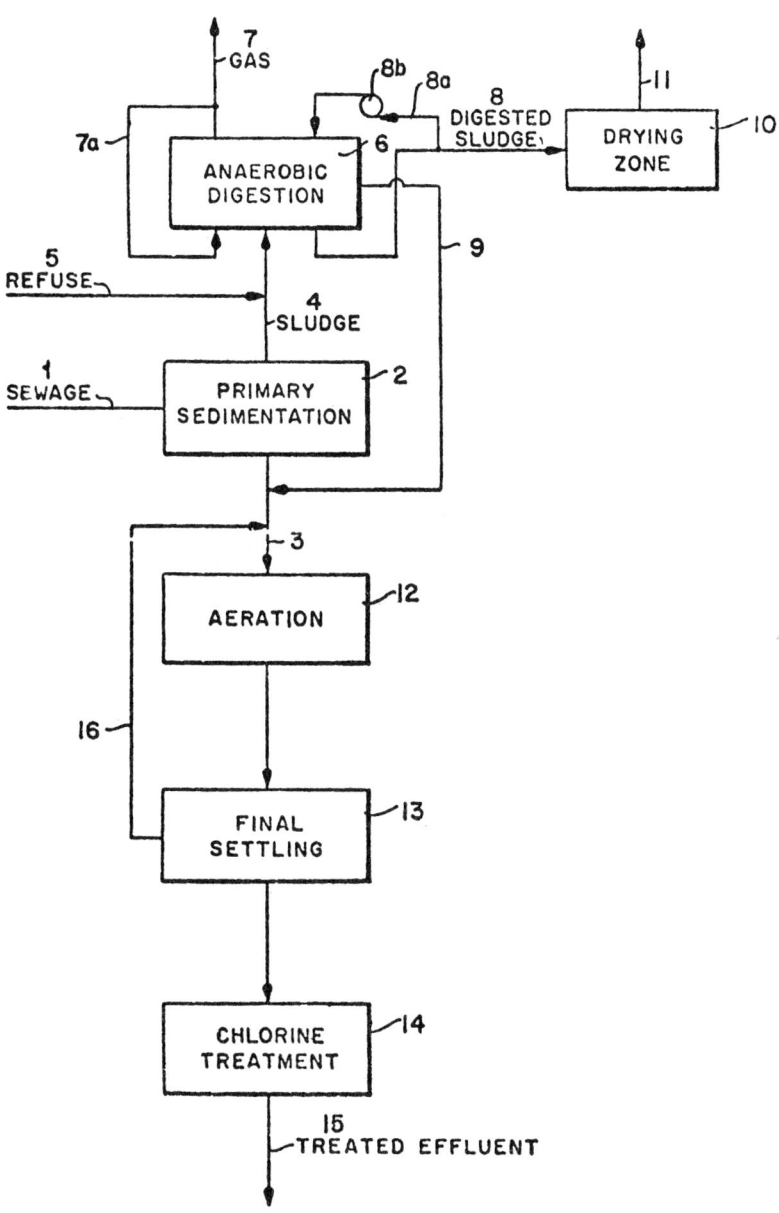

Source: PB 220 821

otherwise comminuted refuse 5 is then introduced into the sludge 4, and the mixture enters anaerobic digestion zone 6, e.g., a substantially closed container having a blanket of oxygen-free gas. Although digestion can take place at 60° to 120°F, the preferred temperature is about 95° to 100°F.

From a practical standpoint, digestion is carried out at pressures slightly above atmospheric, i.e., about 1.0 to 1.5 atmospheres. However, a more efficient operation can be achieved at higher pressures, i.e., about 2 to 10 atmospheres.

Product gas rich in methane (65 to 85%) leaves digester zone 6 through conduit 7, while digested sludge exits at conduit 8, and supernatant liquor is removed by conduit 9 to be mixed with supernatant liquor 3 from zone 2. The product gas can be stored or directly employed as fuel.

Recirculation of the digester contents provides more intimate contact and more complete digestion. Accordingly, conduit 7a recycles a portion of the product gas through the bottom of digester 6, while conduit 8a and pump 8b recirculate some digested sludge to the top. Ultimately, digested sludge is dried in zone 10 to produce low grade fertilizer or soil conditioner 11.

In the prior art manner supernatant liquor 3 can be passed through aeration zone 12, final settling zone 13, and chlorine treatment zone 14. Treated effluent exits from zone 14 through conduit 15. Activated sludge 16 from settling zone 13 is recycled to the supernatant liquor 3. The following examples illustrate the synergistic production of methane from refuse and sludge.

Example 1: In test 1, 50 grams of glass- and metal-free shredded municipal refuse taken from a batch of refuse was mixed with 2 liters of tap water and placed in an anaerobic digester, i.e., a helium-filled, stoppered flask. The flask was maintained at 95° to 100°F by a water bath. In test 2, 2 liters of municipal sewage sludge taken from a batch of sludge were digested in the apparatus of test 1, while in test 3, 2 liters of sewage sludge taken from the same batch as test 2 were mixed with 50 grams of refuse taken from the same batch as test 1, and the mixture was digested in the apparatus of test 1. The results of these three tests are shown in Table 3.9 below.

TABLE 3.9: ANALYSES OF REFUSE AND SLUDGE DIGESTIONS

	Digestion Time (hr)	Gas Produced (ml)	Gas Analysis (%)		
			N_2	CH_4	CO_2
Test 1					
50 grams garbage plus	24	70	–	–	–
2 liters of tapwater	72	70	–	–	–
	264	255	–	–	–
	336	400	–	–	–
Test 2					
2 liters digester sew-	24	945	10.9	50.0	39.1
age sludge	72	1,690	9.2	67.2	23.6
	264	2,495	6.3	74.0	19.7
	336	2,735	–	–	–

(continued)

TABLE 3.9: (continued)

	Digestion Time (hr)	Gas Produced (ml)	Gas Analysis (%)		
			N_2	CH_4	CO_2
Test 3					
2 liters digester sewage sludge plus 50 grams of garbage	24	2,080	1.9	63.0	35.1
	72	4,375	1.6	74.4	24.0
	264	6,230	0.4	82.7	16.9
	336	6,620	–	–	–

Source: PB 220 821

As can be seen from the table, combined sewage sludge and refuse (test 3) resulted in a 50% increase in gas production in comparison to the additive results of tests 1 and 2. Furthermore, the gas produced by test 3 was substantially richer in methane than the gas of test 1 or 2.

Example 2: A second series of comparative tests were run on another batch of refuse and sludge, and the results are shown in Table 3.10 below.

TABLE 3.10: COMPARATIVE TESTS

	Digestion Time (hr)	Gas Produced (ml)	Gas Analysis (%)		
			N_2	CH_4	CO_2
Test 1					
50 grams garbage plus 2 liters of tapwater	96	275	36.8	34.7	28.5
	192	330	–	–	–
	432	1,510	13.7	65.0	21.3
Test 2					
2 liters of digester sewage sludge	96	1,775	11.7	57.7	30.6
	192	2,245	16.1	62.7	21.2
	432	2,780	12.5	69.2	18.3
Test 3					
2 liters digester sewage sludge plus 50 grams of garbage	96	4,020	6.9	68.0	25.1
	192	4,840	5.4	77.8	16.8
	432	6,080	6.7	79.8	13.5

Source: PB 220 821

MANAGEMENT ASPECTS OF REFUSE BIOCONVERSION

At the Bioconversion Energy Research Conference of June 1973, P. Meier of Curran Associates, Inc. discussed the management aspects of refuse bioconversion. He said that if the bioconversion of solid wastes to energy is ever to progress to actual implementation, it is important that academic research into the feasibility of bioconversion processes bear in mind the real world conditions under which the process would function. Thus, the search for optimal bioconversion conditions should not ignore the trade-offs with other unit operations and management considerations that will constitute a working system. Factors to be considered in a total systems evaluation include the subsequent handling and disposal of the nonbioconverted solids fraction, the handling and marketing of the energy resource itself, the preparation of solid waste prior to bioconversion, the location of the bioconversion facility, and the objectives from a solid waste management viewpoint. From a strictly practical viewpoint, energy bioconversion is not only a potential contributor toward mitigation of the national energy crisis, but also a potential solution to the municipal refuse disposal problem. However, even if the objectives of the latter are held subordinate to maximizing energy production, (and it should be obvious that the two sets of objectives might be in conflict), there are numerous trade-offs to be considered between the bioconversion process itself and its precedent and subsequent unit operations needed for practical implementation. It should be noted that although this paper addressed itself to the particular problems of anaerobic digestion of municipal refuse to yield methane, the principles elaborated herein apply equally to other potential energy bioconversion opportunities.

Precedent unit operations include size reduction of refuse, and the removal of inert fractions such as metals and glass which would otherwise needlessly consume digester capacity and hinder mixing. Insofar as the digestion process is a wet process, wet size reduction would seem preferable over dry size reduction. Wet size reduction processes have advantages of lower power requirements, lower noise levels, a more homogeneous effluent and greater efficiency, and devices developed by the pulp and paper industry appear particularly suited to this purpose.

Modern hydropulpers can be operated at a wide range of solids concentrations and have the additional advantage of automatically ejecting metals and nonpulpable items from the pulping operation. One currently marketed device even has the ability to reject glass bottles without breakage or shattering. Municipal refuse, especially residential refuse with little commercial refuse, may contain up to 33% by weight of metals, glass, ceramics and ash; these constituents not only occupy valuable digester space (the capital costs of digesters being largely a function of volume) but would also cause excessive wear and tear on pumps and appurtenant devices.

Because a key operating variable of the digestion process is solids concentration, precedent operations must therefore be consistent with the required feed conditions. For example, if the size reduction process is of the pulping variety, then the solids concentration of the digester feed may be limited by the maximum concentration that can be pumped.

Several investigators have analyzed the optimum proportions of sewage sludge and refuse. However, if energy recovery is also to make any substantial contribution to the sewage sludge and municipal refuse management problems, these proportions are more or less fixed; if successful digestion is not possible at the given sewage sludge to refuse ratio then there will still be a surplus of refuse to be disposed of by other means, and from a solid waste management viewpoint energy bioconversion will lose much of its economic appeal if a second disposal means must be found for the surplus refuse. Research should therefore be focused on getting bioconversion to function at the fixed proportions of refuse and sewage sludge expected by a municipal system; some sacrifice of energy yield or the need for nutrient additions may well be offset by the cost advantages of being able to achieve total integrated conversion. Regardless of the conditions under which the anaerobic digester is operated, any full scale working facility will be faced with the problems and costs of sludge disposal. Dewatering and disposal of sludges is an expensive operation even under the most favorable circumstances. Fortunately, the existing experience with refuse sludges, although still very limited, indicates much better dewatering characteristics than sewage treatment plant sludges of comparable solids concentrations. For example, it has been found that vacuum filtration or centrifugation of a refuse sludge does not require any ferric chloride or polymer conditioners for optimum dewatering performance. Nevertheless, it is desirable that the solids concentration of the digester effluent be as high as possible. Indeed, the most important trade-off in a working facility will be between the desirability of a high solids concentration so as to minimize subsequent sludge handling costs and the desirability of a relatively lower solids concentration in the digester so as to minimize mixing and pumping costs.

Management of the energy resource itself may be crucial to the success of energy recovery of solid waste. Technical problems include the presence of gases that are highly corrosive to generating equipment and gas engines (which must therefore be removed by subsequent unit operations or by sacrificing some of the potential energy yield in order to minimize hydrogen sulfide production), and marketing and distribution of the recovered energy may be hindered or rendered uneconomic by a variety of locational and institutional problems.

One of the more promising ideas for municipalities that already have a municipal power utility is to generate their own peak power using digester fueled gas turbine units at the location of the digester facility. Many small municipal utilities presently face problems of obtaining adequate power supplies from the large investor-owned utilities; but since the municipal utilities already own and operate distribution systems, a major obstacle to the marketing and distribution of bioconverted energy is overcome.

That this potential for on-site power generation is not academic is demonstrated by the Orange County Sanitation District in California, which built a digester gas turbine as early as 1966. The district produced 2.3 million cubic feet of digester gas per day, most of which was previously wasted. Although the unit is very small (1 Mw) and intended primarily as a source of stand-by energy during public power outages, the technical feasibility of the concept has nevertheless been demonstrated.

A total systems evaluation of a concept that embodies both energy recovery and new methods for refuse collection has been under study by Curran Associates for the EPA National Environmental Research Center, Cincinnati, Ohio. Although prime emphasis is on solid waste management aspects rather than energy recovery per se, the concept calls for residential refuse to be ground in the individual home in appliance-like refuse grinders, transport of the ground refuse in the sanitary sewer system, and joint treatment of ground refuse and sewage at modified conventional treatment facilities that include anaerobic digesters for energy recovery. There are numerous technical and economic questions to be resolved, especially potential deposition problems in the sewer system and the high cost of the grinder designs.

REFERENCES

(1) Wise, D.L., et al, *Fuel Gas Production from Solid Waste,* Report No. 1151, Dynatech Corporation, Cambridge, Massachusetts, January 1974.

(2) Cooney, C.L., Lindsey, E.E., Kirk, R.S. and Oyewole, S., *Fuel Gas Production from Solid Waste,* Supplemental Report, Subcontract Programs, Dynatech R/D Company Report 1213 to National Science Foundation, July 1974.
(3) Pfeffer, J.T., *Reclamation of Energy from Organic Refuse,* Final Report Grant No. EPA-R-80076, Dept. of Civil Engr., University of Illinois, Urbana, 1973.
(4) Andrews, J.F., "A Mathematical Model for the Continuous Culture of Microorganisms Utilizing Inhibitory Substrates," *Biotechnology and Bioengineering,* 10, 707, 1968.
(5) Andrews, J.F., "Dynamic Model of the Anaerobic Digestion Process," *Journal of the Sanitary Engineering Division, Proceedings of the American Society of Civil Engineers,* 95, SA1, 95, 1969.

URBAN TRASH METHANATION–POCE

The source of the material in this chapter is the report, PB 240 768. For a complete bibliography, see page 222. As a result of the work of Pfeffer and the Dynatech Corporation (described in the previous chapter), the National Science Foundation decided to undertake a proof-of-concept experiment (POCE) to determine the economic and technical viabilities of large-scale conversion of municipal solid waste (MSW) to methane. The MITRE Corporation prepared the following report for NSF which presented assessments and recommendations for each of the technologies contributing to the urban trash methanation (UTM) process. The report, therefore, provides an overall view of the current state-of-the-art of solid waste processing, anaerobic digestion, gas scrubbing and sludge disposal.

FEED PREPARATION

Control of Digester Feed

The feed preparation process (or front-end) may be functionally defined as the acceptance of raw urban solid waste with varying composition, texture, and moisture content, and the conversion of this input to a relatively homogeneous feedstock to meet the needs of the methane-producing digestion process. Variations in the feed preparation process may have significant effects on digester feedstock and, ultimately, on the overall efficiency of the UTM facility. Specifically, front-end design variations will affect four significant characteristics of the digester feedstock: (1) the percentage of the digestible organic fraction in the input solid waste stream that will ultimately reach the digesters; (2) the ratio of digestible to nondigestible materials in the digester feedstock; (3) the size of the particles in the digester feedstock; and (4) the potential presence of digestion-inhibiting substances in the feedstock.

Due to the impact that the feed preparation process may have on overall UTM plant efficiency, it would seem logical to include a versatile front-end. However, the high cost of such a front-end and the successful demonstration of similar facilities designed for preparation of the feedstock for combustion, resource recovery, and pyrolysis may obviate the need for a versatile front-end in the UTM facility. For example, the St. Louis trash separation facility, which is preparing feed for burning in a power plant boiler, is retrieving a fraction that is approximately 87% of the total milled input (by weight) and is approximately 96% organic. Similar data have been projected for the National Center for Resource Recovery (NCRR) demonstration plant that is being constructed in New Orleans (a fraction that is approximately 55% of fhe total input by weight and is approximately 90% organic).

Due to the specific requirements being set by NCRR in the design of the system components (based on observations by NCRR at its other plants), these projections are given with a relative degree of confidence. Size reduction and separation requirements for the methane fermentation process may be as extensive or demanding as that provided in the above facilities.

Thus, it may be more cost-effective to link the UTM facility to an existing front-end rather than to build a new one. Should this approach be taken, explicit provisions for the determination of the capital and operating costs of this front-end relative to the preparation of the organic fraction required for digester feedstock and any by-product credits and/or waste disposal should be made.

If it is not feasible to link the facility to an existing front-end, a simplified feed preparation facility may be designed, based on the successful results of experiments to date. Such a front-end is described later in this section.

With respect to particle size variations to optimize gas production in the digester, an existing front-end may be fitted with a secondary shredder to provide this capability, if needed. However, adequate mixing may have a pulping effect on most of the digestible organics. This pulping effect is believed to occur within a few hours after the addition of a batch of feedstock into the digester. Thus, it may be feasible to initiate the process without a secondary shredder, test for the occurrence of this pulping effect, and provide a contingency plan for addition of a secondary shredder should the pulping effect not occur as expected.

The fourth digester feedstock characteristic that may be controlled by front-end design is the presence of digestion-inhibiting substances. Highly toxic substances (e.g., pesticides or other commercial poisons) may be present in the municipal waste stream. However, the likelihood of these substances occurring in concentrations large enough to slow or halt the digestion process is quite slim. This question of concentrations of toxic substances is discussed further in the section on anaerobic digestion. Monitoring for such substances would require very extensive and elaborate equipment and techniques that are not justifiable at this time.

Current Feed Preparation Experiments

Particle size is an important variable not only in digester efficiency, but also in determining separation efficiency. In addition, separation efficiency can be affected by variations in separation equipment, machine operation, and feedstock characteristics.

Laboratory tests conducted for EPA at Stanford Research Institute (SRI) demonstrated that there would be no problem in the commercial application of air classifiers in the processing of paper-containing waste, particularly for metal and glass removal; removal of nonbiodegradable material; and pretreatment for composting, fermentation, or retorting, with maximum particle sizes of one to two inches. The removal of nonbiodegradable fines from dry waste was best achieved with a combination of screening and air classification. A similarity was found in the aerodynamic characteristics of the fine dust and grit and the paper constituents of the sample, thus making the screening necessary.

However, difficulties arose in the air classification of the coarser, paper-containing fractions after screening had taken place because of the small differences in density of the remaining fines, paper, cardboard and plastic, and the intimate mixture produced by both the shredding and screening operations.

The use of conventional hammer mills, which tend to tear rather than shred fibrous material, resulted in a dry pulping, or felting, effect. This caused the shredded material to agglomerate, forming a floc of paper and cardboard that picks up and carries with it a great deal of other light and fine material. This felting was seen as a problem to laboratory-scale air classifiers.

The column throat and, thus, the particle size used in a commercial unit would be much larger than that used in the laboratory and felting effects can be minimized by increased throat velocity. A simpler solution was also suggested. It was felt that if the refuse were slightly wetted, the aerodynamic characteristics of the fine dust and grit would be sufficiently differentiated from the paper constituents, and that the fines (including light plastics) would be carried off at a lower throat velocity than the paper. Subsequent experiments confirmed the expected results of wetting.

Separation of ferrous metals and air classification of nonbiodegradable fines from biodegradable materials are more efficient with larger particle sizes. However, one other critical separation (or nonseparation) step needs to be considered. Most food waste, heavy plastic, yard waste, cloth, and wood are heavier than paper and are separated from the light organic fraction during air classification and appear in the rejected heavy fraction. Ideally, since food waste, yard waste, and wood, which are biodegradable, and cloth, which is partially biodegradable, make up 17 to 32% of all municipal solid waste, most of this should be included in the digester feed fraction. Ideally, plastics should be excluded, but this fraction is usually so low—trace to 3%—that it would have little effect on digestion (1). Thus, in this case, less separation efficiency is desired.

It was discovered by SRI that to achieve "greater" separation efficiency of these heavier organics, larger particle sizes were required (2). To reverse this effect, smaller sizes would be desirable. For the purpose of producing a digester feed, particle sizes would have to be large enough to counteract the felting effects, yet small enough to allow for the inclusion of food wastes, yard wastes, and wood in the product from the second stage of classification.

MITRE has completed a testing project for St. Louis in which the refuse shredding and air classification systems were used for separating the organic fraction from urban solid wastes. The resulting organic fraction was used as supplementary fuel in a utility boiler.

The classifier evaluated was a Radar Air Classifier system. Tests were run using various configurations of throat size and reject angle. Other parameters that varied during the test program include the grate bar size and hammer condition of the mill, the feed rate, air velocity through the classifier, type of material and moisture content, and particle size.

Due to a defect in design, the grate bar size had little effect on particle size; however, the carryover of glass and metal from the milled unclassified to the light fraction was investigated on a particle-size basis. Generally, it was found that carryover increases as particle size decreases, implying that classification is improved for larger particle sizes.

It was concluded that feed rate should be set according to classifier throat size and periodically adjusted for changing density in the trash. Experiments showed that less than a 9% inorganic component of the light fraction was obtained at feed rates varying from 48 to 60 tph. Best results were obtained with a throat size of $23^1/_2$ inches and feed rates varying from 53 to 60 tph.

Throat velocities in the separation zone varied from 2,400 to 2,700 fpm for the 20-inch throat configurations and from 2,300 to 2,600 fpm for a $23^1/_2$-inch throat. The velocity range may vary without severely affecting efficiency because of the differences in density between the light materials, such as paper, and the heavies, such as glass, metal and rock.

Moisture had an adverse effect on classification. The added weight was not considered a factor since the density of glass and metal is still considerably greater than paper. The problem was that the wet paper was much more cohesive and tended to trap the inorganic fines. Grass also had a similar effect.

The material balance was determined in the late spring of 1974. On the milled, unclassified material, approximately 87% (by weight) was accepted as the light fraction and approximately 13% was deposited as the heavy fraction.

The composition of the light fraction for each of the seven classifier configurations is summarized in Table 4.1. For the most part, the remaining light fraction cited in the table is totally digestible. It may, however, contain a small amount of plastic, rubber or synthetic fibers.

It was concluded that the larger the particle and the slower the feed rate, the better the air classifier separates the material. The large particles have less of a tendency to be entrapped by the lighter material and to be forced into the light fraction. The slower feed rate allows particles to enter the separation zone more evenly and reduces the clumping of wet material, thereby increasing classification efficiency.

Thus, at least for urban solid waste of a composition similar to that of St. Louis, the technology of separation of organic fraction from inorganic fraction through air classification has been demonstrated. Magnetic separation of ferrous metals is also fairly well understood. It has been noted that large items such as appliances need to be reduced in size to pieces about 8 inches or smaller in order to be processed by a magnetic separator (3). However, while reduction to an extremely small size is not necessary, it is important that particles be physically freed from each other. Reduction to extremely small sizes may retard separation by causing intimate mixing of particles.

TABLE 4.1: LIGHT FRACTION COMPOSITION BY PERCENT WEIGHT— RADER AIR CLASSIFIER

	VOLUME FLOW = 29 x 10³ to 31 x 10³ CFM THROAT VELOCITY = 2400-2700 ft/min				VOLUME FLOW = 32 x 10³ to 33.5 x 10³ CFM THROAT VELOCITY = 2300-2600 ft/min			
Throat Size	20"	20"	20"	20"	23½"	23½"	23½"	23½"
Configuration	Straight Column	Slant To Left	Zigzag	Straight Column	Straight Column	Straight Column, Pinched To 19" at Bottom	Straight Column	Straight Column
COMPONENT								
MAGNETIC	---	---	1.9	1.2	.7	.7	.8	.5
TOTAL METAL	1.9	4.4	2.1	1.8	1.1	1.7	1.2	1.0
GLASS	6.4	3.6	5.0	4.3	3.8	2.8	4.4	2.8
GLASS & METAL	8.3	8.0	7.1	6.1	4.9	4.5	5.6	3.8
REMAINING LIGHT FRACTION (Nearly totally digestible)	91.7	92.0	92.9	93.9	95.1	95.5	94.4	96.2

Source: PB 240 768

Feed Preparation Options

Two options seem apparent for front-end design for a UTM facility: to build a front-end that meets only the barest essentials in producing the feed needed for digester experiments, or to use feed produced by some other existing or soon-to-be-completed resource recovery system. The first option, while limited in terms of the possible variations in feed composition, is less expensive than building a complete separation facility. Moreover, there is the convenience of having the front-end located contiguous with the rest of the system. The second option, while possibly presenting constraints regarding plant location, has much lower capital costs and, assuming a versatile separation facility is used, the capability of producing a high-quality feed.

Option 1—Basic Front-End: Considering the problems in separation efficiency cited in the SRI study, the most efficient process operation would seem to be one in which screening and secondary milling follow air classification. In this manner, flocculation of fines could be minimized. If ferrous metals are to be recovered, it would probably be more efficient to put a magnetic separator before the air classifier. This would remove some of the constituents of the heavy fraction, making air classification easier.

A second magnetic separation of the air classifier heavies would recover an additional small amount of ferrous metals. After air classification, the light fraction should contain mostly paper, cardboard and light plastic. If the particle sizes are small enough there should also be a significant amount of food waste, heavy plastic, yard waste, cloth, and wood in the light fraction. This light fraction should then be screened to remove the remaining fines. The system can be expected to produce a material balance similar to that shown in Figure 4.1. Using this system, the ultimate feedstock particle size can be adjusted (for digester experiments) by adjusting the secondary mill, although this mill is optional and should be omitted pending investigation of pulping effects of the solids in the digester.

The basic front-end of Option 1 is heavily dependent upon the efficiency of the air classifier used. Generally, air classifiers have been successful in separating a large part of the organic fraction of milled solid waste. The nondigestible fine particles that stay with the light (mostly organic) fraction from the air classifier are easily separated by the screening. Those nondigestibles that get through the classifier and are not fine enough to fall through the screen are a small fraction of film plastics.

Figure 4.2 presents a simplified block diagram of the components of Option 1. Installed equipment costs for this Option (see Table 4.2), are estimated to be about $869,000 (1973 dollars), based on the component cost estimates used by Dynatech.

Option 2—The Use of an Existing Front-End: According to National Center For Resource Recovery (NCRR), at least 25 communities were operating shredders by 1976 and each of these sites represented a potential for additional processing units. Several separation and recovery systems are already in operation, including St. Louis, Mo., New Castle, Del., and Franklin, Ohio. A substantial savings in capital costs can be realized if an existing or soon-to-be-completed front-end is used in lieu of construction of this part of the POCE facility. However, capital and operating costs of this front-end must be analyzed very closely and the possibility of ultimate commercial application based on data from this existing facility must be considered in terms of the specific requirements for digester feed stock preparation.

Presently, the State of Delaware is operating a 500-tpd facility. This process, located in New Castle County, uses air classification, magnetic separation, screening, rising current, heavy media, and electrostatic separation as well as optical methods for separating municipal solid waste into paper, ferrous and nonferrous metals, glass, and organic fractions. In addition to the regular markets for the glass, paper and metals, negotiations are under way between the contractor and Delmarva Power and Light Company to process the refuse-derived organic fuel in existing oil-fired utility boilers. A similar system, designed by NCRR, has been installed in New Orleans.

The most important aspect to consider in the use of an existing front end is the composition of the organic fraction that can be made available to the UTM plant. A plant designed for organic separation should be capable of producing the high-quality organic reported above for St. Louis. Assuming the existence of a market for the organic fraction separated by an existing separation facility, the POCE may have to pay for digester feed and feed transport. If no local market exists, acceptance of feed by the UTM facility would decrease the separation facility's disposal costs and likely result in free digester feed.

The POCE plant interface requirements for Option 2 are illustrated in Figure 4.3, and installed equipment cost estimates in 1973 dollars are shown in Table 4.3. In summary, it would seem most advantageous to construct a UTM facility in

FIGURE 4.1: MATERIAL BALANCE FOR 15 TONS PER HOUR OF MUNICIPAL SOLID WASTE

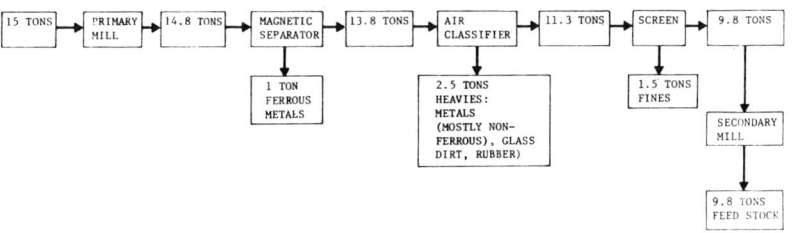

FIGURE 4.2: OPTION 1, BASIC FRONT-END

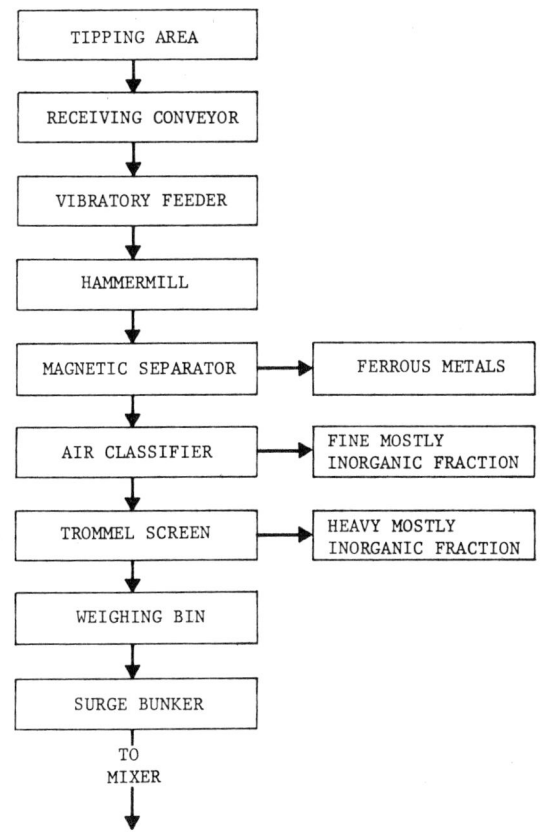

Source: PB 240 768

TABLE 4.2: INSTALLED EQUIPMENT COSTS, OPTION 1

ITEM	COST
Hammermill	$ 50,000
Magnetic Separator	10,000
Trommel Screen	21,000
Air Classifier	190,000
Scales	55,000
Surge Bunkers and Weigh Bins	320,000
Vibratory Conveyors	40,000
Belt Conveyors	103,000
Sub total	789,000
Miscellaneous Equipment	80,000
TOTAL	$869,000

FIGURE 4.3: OPTION 2, ABBREVIATED FRONT-END

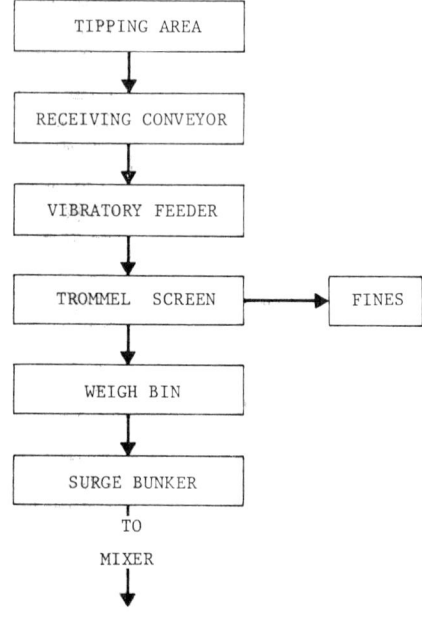

Source: PB 240 768

TABLE 4.3: INSTALLED EQUIPMENT COSTS, OPTION 2

ITEM	COST
Vibratory Conveyor	$ 40,000
Trommel Screen	21,000
Scales	55,000
Surge Bunkers and Weigh Bins	320,000
Belt Conveyors	41,000
Sub Total	477,000
Miscellaneous Equipment	80,000
TOTAL	$557,000

Source: PB 240 768

conjunction with an operating front-end facility, preferably one that has no market for its organic fraction and preferably one with some experimental capability, as would be the case in a demonstration plant. The National Center for Resource Recovery, New Orleans plant is an example of such a facility.

ANAEROBIC DIGESTION

The Microbial Process

The anaerobic digestion of organic compounds to methane and carbohydrates can be regarded as a two-stage process. In the first stage, acid-forming bacteria break down complex organic molecules such as cellulose, fats, proteins and carbohydrates to low molecular weight volatile acids, such as acetic, butyric and propionic. In the second stage, methane-forming bacteria convert the volatile acids to methane and carbon dioxide.

The acid-forming bacteria strains are a more heterogeneous group, faster growing and less sensitive to changes in their environmental conditions than methane-forming bacteria. Though acid formers are anaerobic, many strains are facultative anaerobes, having some tolerance to oxygen. Oxygen is fatal to methane formers which are obligate anaerobes.

Some anaerobic acid-forming bacteria and the materials which they digest are shown in Table 4.4 (4). The cellulose digesters will be most important in converting urban solid waste to methane, since the digestible organic fraction of solid waste is almost entirely cellulose. Some methane-forming strains and the substrates they metabolize are given in Table 4.5. A proper balance between the acid-forming and methane-forming bacteria must be maintained.

TABLE 4.4: NONMETHANOGENIC BACTERIA ISOLATED FROM ANAEROBIC DIGESTERS

BACTERIUM / ISOLATED ON	CELLULOSE	STARCH	PROTEIN PEPTONE	PROTEIN CASEIN	LIPID
Aerobacter aerogenes					
Alcaligenes bookerii					X
A. faecalis	X				
Bacillus sp					
B. cereus var. mycoides		X		X	
B. cereus	X	X	X	X	
B. circulans			X		
B. firmus			X		
B. knelfelhampi					
B. megaterium	X	X		X	X
B. pumilis			X	X	
B. sphaericus			X	X	X
B. subtilis			X	X	X
Clostriduim carnofoetidum	X				
Escherichia coli			X	X	
E. intermedia					
Micrococcus candidus		X			
M. luteus					X
M. varians		X	X	X	
M. ureae		X			
Paracolobactrum intermedium			X		
P. coliforme			X		
Proteus vulgaris	X				
Pseudomonas aeruginosa	X				
P. ambigua					
P. oleovorans					X
P. perolens					X
P. pseudomallei					
P. reptilivora	X				
P. riboflavina	X				X
P. spp.	X	X	X	X	X
Sarcina cooksonii					
Streptomyces bikiniensis					X

Source: PB 240 768

In a balanced process, methane producers fully utilize the acids produced in the first stage of digestion. Since methane-forming bacteria are slow to multiply and very sensitive to environmental stress, any adverse change in environmental conditions will inhibit growth of methane producers and disrupt the balance between the two groups. An excess of volatile acids is usually the best indicator that the anaerobic digestion process is operating at reduced efficiency and is in danger of falling. Once volatile acid concentrations exceed 2,000 ppm, the process begins to "sour." Growth of a microbial population follows a geometric progression with time. Individual cells grow and divide at approximately the same rate as their parent cells, so that the number of cells doubles during each generation.

TABLE 4.5: METHANOGENIC BACTERIA

BACTERIUM	SUBSTRATES	PRODUCTS
Methanobacterium formicum	CO $H_2 + CO_2$ formate	CH_4
M. mobilis	$H_2 + CO_2$ formate	CH_4
M. propionicum	propionate	CO_2 + acetate*
M. ruminantium	formate $H_2 + CO_2$	CH_4
M. soehngenii	acetate butyrate	$CH_4 + CO_2$
M. suboxydans	caproate & butyrate	propionate & acetate*
Methanococcus mazei	acetate & butyrate	$CH_4 + CO_2$
M. vannielii	$H_2 + CO_2$ formate	CH_4
Methanosarcina barkeri	$H_2 + CO_2$ Menthanol acetate	CH_4 CH_4 $CH_4 + CO_2$
M. methanica	acetate butyrate	$CH_4 + CO_2$

*Acetate or propionate converted to CH_4 in a two-step process.

Source: PB 240 768

A typical growth curve for a bacterial population is shown in Figure 4.4. Before a constant growth rate is established, a brief period of time, the lag phase, usually elapses. After an exponential growth period, as nutrients become exhausted and toxic metabolites accumulate, the maximum stationary phase of the culture is established. In this phase, the number of viable bacteria appear to remain constant at a maximum value.

The goal in anaerobic digestion will be to maintain the rate-limiting bacteria (the methane producers) in the maximum stationary phase as long as possible. The stationary phase is usually a statistical phenomena, since growth of the population continues but is counterbalanced by death which occurs at an equivalent rate. When the death rate begins to exceed the growth rate, the culture enters the death phase. After a constant death rate is established, the culture dies exponentially.

FIGURE 4.4: BACTERIAL GROWTH CURVE

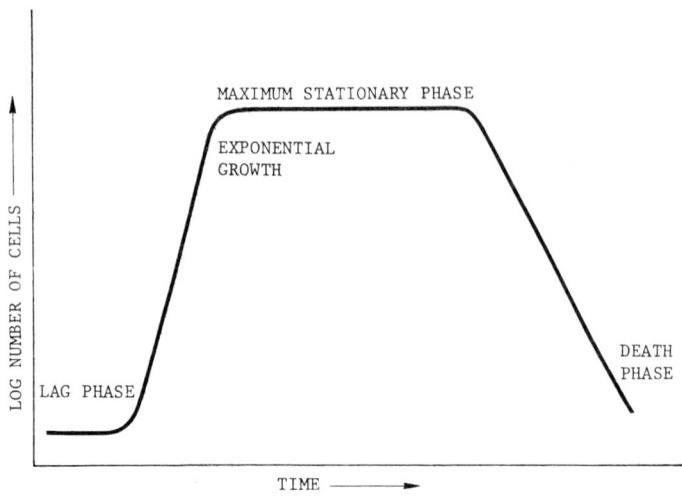

Source: PB 240 768

Maximum gas production occurs when methane-producing bacteria have reached the maximum stationary phase. Environmental conditions should be manipulated to keep them in this stage and forestall bacterial death. The time required to reach maximum stationary growth at startup or after a change in environmental conditions needs to be considered in scheduling experiments and in returning the digester to operating conditions after a breakdown in plant operation.

Using a seed culture, steady state may be achieved in two to three months, where steady state is defined as maintaining variations within 20% of the maximum gas production. This is essentially independent of digester size, as the seed culture would be appropriately scaled. Without a seed culture much larger growth periods may be required.

Pfeffer and Liebman (6) developed a seed culture for thermophilic operation using pulped bond and newspaper plus raw sewage sludge and nutrients. Cooney (7) used a continuous culture technique to select a stable thermophilic microbial population capable of converting cellulose to methane gas. He considered such selective culturing of thermophilic methanogenic bacteria to be a potentially powerful tool for facilitating thermophilic digestion of waste.

Gas Production

Anaerobic digestion of carbohydrates, such as cellulose, produces an equal number of mols of CH_4 and CO_2:

$$x(C_6H_{10}O_5) + xH_2O \longrightarrow x(C_6H_{12}O_6) \longrightarrow 3xCH_4 + 3xCO_2$$

The methane, being insoluble, is found in the gaseous phase. The carbon dioxide, on the other hand, is partially dissolved in the water medium. Small quantities of NH_3, H_2S, H_2, N_2 and O_2 will also be found in the gas from anaerobic digestion of organic material.

Yields of gas produced by several investigators from anaerobic digestion of solid waste are shown in Table 4.6. Yields ranged from 6.5 to 10 cubic feet per pound of dry organic matter fed to the digester. This gas has approximately 60% CH_4 and 40% CO_2 on a dry gas basis.

TABLE 4.6: GAS PRODUCTION CHARACTERISTICS

Study	Volume Percent CO_2	Volume Percent CH_4	Yield, scf Gas/lb Volatile Solids Added
Pfeffer (8)	45	55	6.5-7
Christopher (5)	40	60	6.5
Golueke (12)	40-45	55-60	7
Wise (13)	40	60	8-10
Wolf (14)			
Conventional	40	60	9.5
Two-stage	25-30	70-75	10

*The Pfeffer, Golueke, and Wise studies were conducted using MSW or synthetic MSW as feedstock. Feedstock for the other studies was dry dog food. The broad range in results may be attributed to varying operating conditions such as digester temperature, retention time, pH, and alkalinity.

Source: PB 240 768

The composition of gas is controlled primarily by temperature, pH and alkalinity. The retention time will also have some effect since at shorter retention times, more carbon dioxide will be washed out of the system with the replacement of liquid. The CO_2 dilutes the methane, reducing the heating value of the digester gas. Removal of CO_2 can raise the heating value of the digester gas to the 1,000 Btu per standard cubic feet required by natural gas utilities.

The theoretical relationship between gas composition, temperature, pH and alkalinity is illustrated in Figure 4.5 (8). The CO_2 content of the gas increases with higher temperature, higher bicarbonate concentration and lower pH. However, practical limits on buffering would prevent CO_2 concentrations of less than 30 to 35% in the digester gas.

Variables Affecting Anaerobic Digestion

The variables which will have to be controlled during the POCE are listed in Table 4.7. The table also summarizes information on the range over which parameters can be varied and the effect of deviating from the range. Failure to control temperature, pH, toxicity, and volatile acid concentration pose the greatest risk of killing off the methane-forming bacteria. The experimental design for a UTM facility should allow for empirical determination of maximum solids concentration and minimum retention time for optimal methane production.

FIGURE 4.5: THE RELATIONSHIP BETWEEN GAS COMPOSITION, TEMPERATURE, pH, AND ALKALINITY

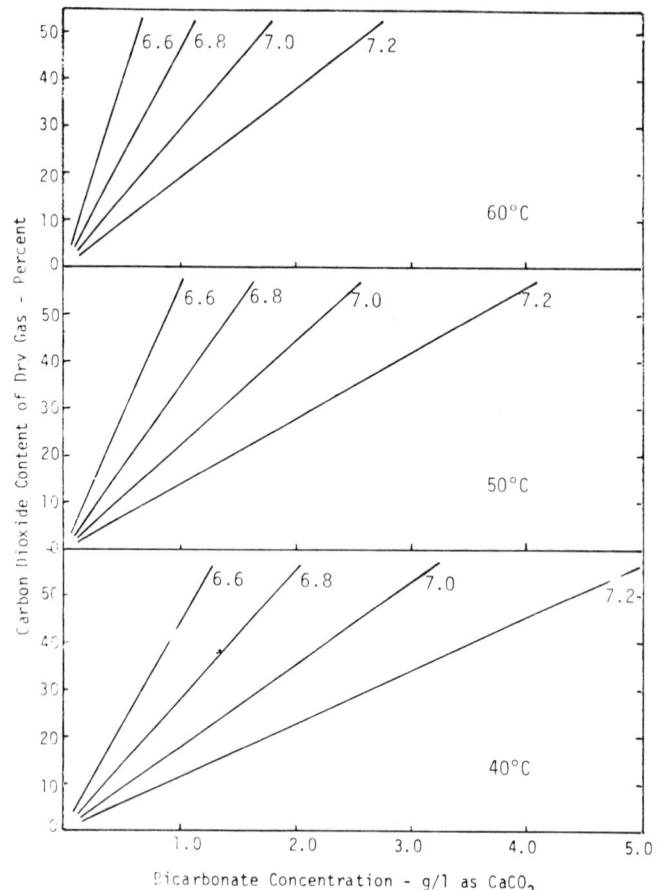

Source: PB 240 768

Temperature: Digestion and gas production can occur over a wide range of temperatures (30° to 60°C) if the temperature is held constant. Once an effective temperature range is established, slight changes in temperature can interfere with the growth of methane producers, upsetting the balance between these bacteria and the acid-producing strains, and resulting in an accumulation of volatile acids.

A plot of gas production versus temperature, Figure 4.6 (10), reveals two peaks of gas production. At about 40°C there is a peak for gas production by mesophilic methane-producing organisms. At about 55°C, there is a peak for gas production by thermophilic methane producers.

TABLE 4.7: DIGESTION VARIABLE CONTROL PARAMETERS

Variable	Limit	Effect of Violating Limits	Effect on Gas Production	Comments
(1) Temperature	30° to 65°C must be held constant	Below 30°C, methane production very slow. Above 65°C, methane bacteria are killed.	Methane production is optimum at 40°C (mesophilic bacteria) and 55°C (thermophilic bacteria).	Fluctuation of temperature can cause microbial imbalance and volatile acid accumulation. As temperature increases, H_2O and CO_2 content of gas also increase.
(2) pH	6.6 to 7.6	pH less than 6.2 is toxic to methane-forming bacteria. At pH above 7.6, ammonia gas can become inhibitory.	Methane production is optimum at a pH of 6.8.	Buffering is required since methane bacteria are sensitive to pH change. As pH increases, bicarbonate alkalinity increases and CO_2 content of gas decreases.
(3) Toxicity	See Table 4.10.	Biological activity is reduced at moderately inhibitory levels and approaches zero as toxic levels are reached.	Methane production is reduced at inhibitory levels and approaches zero at toxic levels.	It is difficult to monitor the input stream for toxic materials. Alkaline metal toxicity can result from addition of caustic for pH control.
(4) Retention Time	Should be minimized experimentally. The minimum must be greater than generation time of methane bacteria removed in liquid and solid effluent.	Below minimum retention time, methane bacteria are washed out before they can reproduce and population falls.	After an initial rapid increase in gas production the daily increment in gas production tapers off with increased retention time.	The relation between the cost of holding digester contents and the value of gas produced will determine optimum retention time. Long retention times increase volume and heating requirements and increase CO_2 content of the gas.

(continued)

TABLE 4.7: (continued)

Variable	Limit	Effect of Violating Limits	Effect on Gas Production	Comments
(5) Required nutrients	Nitrogen Phosphorus Trace quantities of other nutrients	Inadequate N and P will reduce growth of methane bacteria.	Addition of nutrients will improve gas production.	Sewage sludge can provide the materials which are needed in trace quantities for optimal growth and might provide sufficient quantities of N and P.
(6) Bicarbonate Alkalinity	1,500–5,000 mg/l as $CaCO_3$	Below 1,500 mg/l, pH will fall. Above 5,000 mg/l, there is a risk that Na or Ca concentrations can approach inhibitory levels.	As bicarbonate alkalinity increases from 1,500 to 5,000 mg/l, CO_2 content of gas increases.	
(7) Solids concentration	Should be maximized experimentally.	If solids concentration too low, volume and heating requirements will be high. If solids concentration too high, solids won't be kept in suspension. Overloading can cause high volatile acid concentration.	Increasing solids concentration toward optimum will increase methane production. Very high solids concentrations will increase CO_2 content of gas.	Process is most efficient when solids concentration is as high as possible without adverse effects and with consideration of mixing energy requirements.
(8) Volatile acids	50 to 500 mg/l as acetic	Inhibitory to methane producers at concentrations over 2,000 mg/l as acetic.	If volatile acid concentration is increasing, then methane production is decreasing.	Rise in volatile acid concentration, due to microbial imbalance is usually the best indicator that methane-producing bacteria are affected by adverse conditions. High volatile acid concentration is usually symptom, not cause, of process failure.

Source: PB 240 768

FIGURE 4.6: EFFECT OF TEMPERATURE ON GAS PRODUCTION

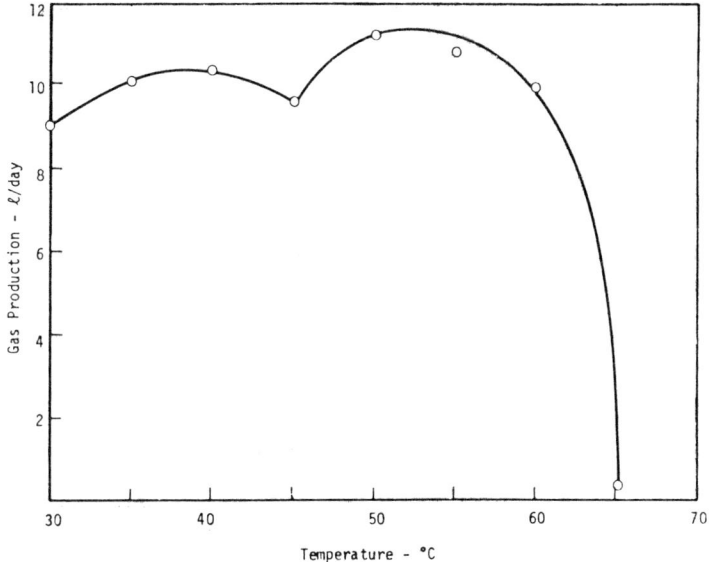

Source: PB 240 768

At temperatures greater than 65°C, methane producers become inactive and methane production decreases. The advantages of operation under either mesophilic or thermophilic conditions are summarized in Table 4.8. At thermophilic temperatures, an increased metabolic rate of microorganisms results in a higher rate of methane production. This makes it possible to use shorter retention times. Cooney (7) found that digestion efficiency (the ratio of organic material converted to gas to organic material fed to the digester) increased at thermophilic temperatures. The conversion of organic material to gas was 67% under mesophilic conditions and 88% under thermophilic conditions.

The major disadvantages of thermophilic operation are higher heating requirements and higher percentage of CO_2 in the digester gas. Additionally, Pfeffer and Liebman (6) observed an accumulation of volatile acids, accompanied by a decrease in methane production, caused by an accidental temperature increase of only 2.5°C in the thermophilic range (from 60° to 62.5°C). After correction, the system took one week to regain its rate of methane production.

Temperature has a significant effect on the bicarbonate equilibrium for a given pH, in that the ratio of bicarbonate to carbon dioxide in solution as carbonic acid increases with increasing temperature. For a pH of 7, at 30°C, this ratio is 4.9 while at 60°C it increases to 5.2. This reduction of CO_2 in solution at increasing temperatures should affect the balance between CO_2 in solution and the gaseous phase, causing a decrease of CO_2 in the digester gas. An increase in temperature, however, decreases the solubility of CO_2 in water.

TABLE 4.8: ADVANTAGES OF MESOPHILIC AND THERMOPHILIC CONDITIONS

MESOPHILIC	THERMOPHILIC
• Less water vapor in gas • Less CO_2 in gas • More types of bacteria grow and produce methane • Lower heating requirements	• Higher rate of digestion and methane production resulting in shorter retention time and smaller digestion volume for the same trash throughout. • Decrease in net sludge formed • Destruction of more pathogenic organisms • Potential for increased digestion efficiency • More rapid regeneration of population after partial souring • Easier maintenance of anaerobic conditions.

Source: PB 240 768

In going from 30° to 60°C, the solubility of CO_2 decreases by a factor of 0.85. This significant reduction in CO_2 solubility, increasing CO_2 in the gas phase, outweighs the slight shift in the CO_2-bicarbonate equilibrium toward increased conversion of CO_2 to bicarbonate. For a given pH, increasing temperature decreases the alkalinity, since the CO_2 lost from the liquid to the gas phase will decrease the bicarbonate concentration.

The effect of temperature on the CO_2 content of the gas was shown in Figure 4.5. At pH 6.8, with a bicarbonate concentration (as $CaCO_3$) of 1 gram per liter, the CO_2 content of the gas increases from about 30% at 40°C to about 50% at 60°C.

The amount of water vapor in the gas phase increases with temperature. The effect of temperature on vapor pressure of water and on the maximum moisture content of the gas at atmospheric pressure is shown in Table 4.9 (14). Thus operating at thermophilic temperatures would increase the cost of drying the product gas.

Anaerobic Conditions: Unlike some acid-forming bacteria, methane formers are strictly anaerobic. Even small amounts of oxygen are detrimental to methane-producing bacteria.

TABLE 4.9: MOISTURE CONTENT OF SATURATED AIR

TEMPERATURE (°C)	MAXIMUM MOISTURE CONTENT (lb H$_2$O/lb air)	VAPOR PRESSURE OF WATER (mm Hg)
30	.027	31.8
35	.036	42.2
40	.048	55.3
45	.064	71.9
50	.086	92.5
55	.114	118.0
60	.151	149.0

Source: PB 240 768

Maintenance of anaerobic conditions must, therefore, be considered in the design of equipment and processes. Special consideration of anaerobiosis should be made during startup. The headspace could be purged with nitrogen or with gas produced in another digester.

A floating lid is often used in sewage sludge digesters to help maintain anaerobic conditions (9). The top of the tank floats on the liquid to accommodate changes in volume. Gases can be piped from the floating lid, but this requires expensive engineering and construction to provide an impeller for mixing and scum dispersal. Additionally, a floating lid may present problems in sealing the digester to prevent loss of gas. Thus, fixed-top tanks may be more appropriate.

pH and Buffering: Anaerobic digestion generally can take place over a range of pH values from 6.2 to 7.6. Most experimental work has shown good gas production at a pH of about 6.8. Slightly higher pH levels may marginally increase gas production; however, the cost associated with extensive buffering likely outweighs the potential benefit.

At pH values below 6.2, acid conditions become so toxic to methane formers that digestion comes to a complete halt. Moreover, methane-producing bacteria are very sensitive to changes in pH. Therefore, pH control by buffering is essential to effective anaerobic digestion.

Some buffering is naturally present from the digestion process but the addition of supplementary buffer is often desirable. The use of lime is widespread in sewage sludge digestion. McCarty (10) recommends NaHCO$_3$. However, Pfeffer (8) tried adding NaHCO$_3$, but it produced an undesirable surge of CO$_2$ and he switched to NaOH. To keep the pH near neutral at a temperature of 40°C requires a bicarbonate alkalinity (as CaCO$_3$) of at least 1.5 grams per liter in the presence of a 30% CO$_2$ atmosphere. At 60°C an alkalinity of at least 1.0 gram per liter is required to maintain the same percent CO_2 and pH level. A higher alkalinity of 3 to 4 grams per liter gives better protection against a drop in pH resulting from excessive volatile acids.

Required Nutrients and Sludge Addition: Sewage sludge usually contains all required nutrients in forms readily available to microorganisms, but solid wastes may not. Solid waste is deficient in nitrogen and phosphorus, and additions of these nutrients would increase microbial growth and gas production (8). The addition of sewage sludge to solid waste improves gas production more than would be expected merely from the increase in organic material. This increase occurred even when N and P were added in adequate amounts to support microbial growth. This increase in gas production is likely due to the content of trace elements and additional organisms in the sewage sludge.

Sludge already actively undergoing anaerobic digestion in a sewage treatment plant should be a richer source of such microbes than raw sewage. Sewage sludge has more buffering capacity than solid waste due to its protein content which can be deaminated by bacteria to provide ammonium ions in solution. Whether the addition of sludge to solid waste provides adequate nutrients depends on the composition of the waste, the sludge, and the ratio of waste to sludge.

It is unlikely however, that the sodium and potassium content of the sewage produced by a given population will be adequate to supply the nutrients needed for digestion of the solid wastes produced by the same population. Thus, supplemental nutrients will be required.

Adequate nitrogen levels for optimal digestion must be maintained without causing ammonia nitrogen levels to reach toxic proportions (8). This can be done with ammonia nitrogen concentration between 100 and 500 milligrams per liter. Pfeffer and Liebman (6) added nitrogen in the form of ammonium chloride to maintain a soluble ammonium nitrogen concentration of 300 to 400 milligrams per liter or 20 grams per pound refuse. They also added four grams of KH_2PO_4 per pound refuse to improve the potassium level. Some elements including sodium, potassium, calcium, magnesium and iron have stimulatory effects at low concentrations, but can be inhibitory at higher concentrations (11).

Toxicity: It is extremely difficult to monitor continuously the incoming waste stream to identify toxic materials before they can interfere with the digestion process. Only materials in solution are toxic to biological life (11). Inhibition or toxicity can be controlled by the inactivation of toxic compounds or by their removal from solution by forming an insoluble complex or precipitate. Additionally, the input waste stream could be diluted below the toxic threshold or an antagonistic material could be added.

Alkali and alkaline-earth metal salts, such as those of sodium, potassium, calcium, or magnesium, are toxic above threshold concentrations (see Table 4.10). The concentration of these salts in municipal wastes is normally low enough to avoid toxicity. Problems can arise, however, when they are introduced at high concentrations for pH control. Low, but soluble, concentrations of heavy metals such as copper, zinc and nickel salts may be quite toxic. This toxicity must be considered in selecting materials for constructing the digestion tank. Precipitating the heavy metals as sulfides removes the toxicity. Sulfide addition must be done with care, since sulfides are corrosive and toxic themselves unless present as insoluble metal precipitates.

TABLE 4.10: TOXICITY LEVELS FOR SELECTED SUBSTANCES

SUBSTANCE	TOLERABLE LIMITS* (mg/ℓ)	TOXIC THRESHOLD** (mg/ℓ)
Na	3500-5500	8000
K	2500-4500	12000
Ca	2500-4500	8000
Mg	1000-1500	3000
NH_3	--	1500-3000 (if pH > 7.4)
Sulfides	--	200
Heavy Metals: Copper, Zinc, Nickel, Chromium	The toxicity depends on the ionic state of the material, its solubility, and possible precipitation as a sulfide.	

*Reference (11).
**Reference (15).

Source: PB 240 768

Even ammonia, which serves as a nitrogen source, can be toxic at sufficiently high concentrations. Ammonia gas in solution is inhibiting at much lower concentrations than ammonium ion. It is, therefore, beneficial to shift the $NH_4^+ \rightarrow NH_3 + H^+$ equilibrium to the left by lowering pH (pH 7.2 or lower).

Ammonia gas becomes inhibitory if the concentration is between 1,500 and 3,000 milligrams per liter and the pH is greater than 7.4 to 7.6. When ammonia-nitrogen concentration exceeds 3,000 milligrams per liter, then the ammonium ion itself becomes quite toxic (16). Sulfides are toxic at concentrations above 200 milligrams per liter.

Many organic compounds may inhibit anaerobic digestion. These include organic solvents, alcohols and long chain fatty acids at high concentrations. Pesticides in the solid waste also present a potential problem.

Feed Solids Concentration: For a fixed gas production, increasing the feed solids concentration in the digester reduces digester volume requirements, digester heating and subsequent dewatering costs; at the same time, the energy required for mixing is increased.

Water requirements for bacterial growth limit the solids concentration to less than 40% by weight. However, it is likely that well before this limit is reached, physical, economic and positive energy balance constraints (that is, energy increase required for greater mixing versus marginal increase in gas energy production) will limit the maximum suspended solids concentration.

Sewage sludge digesters are usually operated at a total solids concentrations from 3 to 10%. Shulze (17) found that sewage sludge solids could be fed to a laboratory scale digester at concentrations up to 37% as long as volatile acid concentration is maintained below 2,000 ppm. Overloading the digester with organic solids increases volatile acid concentrations, CO_2 content of gas, and the time required to attain steady-state digestion.

If the solid waste is dilute and a high retention time is being used, a very large digestion tank would be required with an excessively high heating cost. At very low solids concentrations, long retention times are not economically feasible. In general, municipal solid waste has an advantage over sewage as a substrate for methanation in that solid waste feed can have a greater solids concentration. However, if wet pulp processes are used in the front ends, solid waste may have to be diluted greatly if it has to pass through a series of pumps and separators as a pulp.

Pfeffer and Liebman found that increasing feed solids concentration has no significant effect on gas production per unit dry solids up to 35% dissolved solids. However, because decreased heating costs outweighed the increase in mixing costs, they found that a higher feed solids concentration gave them a greater net energy recovery and a greater net increase in methane production.

Solids Retention Time: For a continuous or intermittent process, the retention time is a function of the solids concentration and the rate of loading of the digestion tank (18). This relationship may be expressed as

$$T = 62.4 \frac{S}{L}$$

where S is the fraction of solids in sludge on a dry weight basis; L is solids loading in pounds per cubic feet per day; and T is the retention period in days.

Increasing retention time increases costs since larger digester volumes and additional heating are required. Cumulative gas production increases asymptotically with time as shown in Pfeffer's plots of gas production versus retention time at mesophilic temperatures [Figure 4.7 (8)] and at thermophilic temperatures [Figure 4.8 (8)]. More gas is produced at a longer retention time, but the daily increment in gas production tapers off.

Data from Figures 4.7 and 4.8 show that at 60°C, 6.5 standard cubic feet (scf) of gas per pound of volatile solids (V.S.) were produced during the first ten days. During the next ten days, an additional 0.4 standard cubic foot per pound volatile solids was produced (a 6% increase) and during the next ten days, only 0.2 standard cubic foot per pound was produced (a 3% increase). Short retention times also appear to give highest gas production per day and a greater washout of dissolved CO_2 resulting in higher methane content in the gas produced. However, short retention times result in a small percent of the volatile solids being converted to gas with an associated increase in required quantity of feed and disposal of wastes. The rate at which materials are withdrawn from the digester will affect the microbial population. As solids retention time decreases, the relative proportion of active cells washed out of the system increases.

If the retention time is too low, microorganisms of anaerobic digestion will be washed out faster than they can reproduce themselves. The minimum retention time, therefore, depends on the temperature and the generation time of the microorganisms. Near the minimum retention time, the efficiency of anaerobic digestion is low and process dependability poor. McCarty (19) recommends a retention time of at least two and one-half times the minimum. The optimum retention time permitting efficient production of methane and to maximize the net energy produced should be determined.

FIGURE 4.7: GAS PRODUCTION AT MESOPHILIC TEMPERATURES

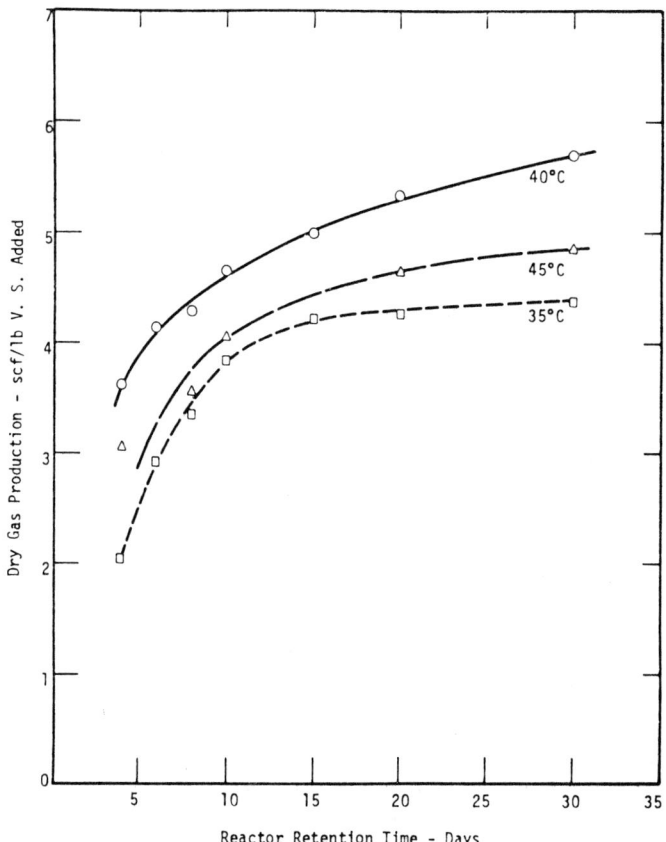

Source: PB 240 768

FIGURE 4.8: GAS PRODUCTION AT THERMOPHILIC TEMPERATURES

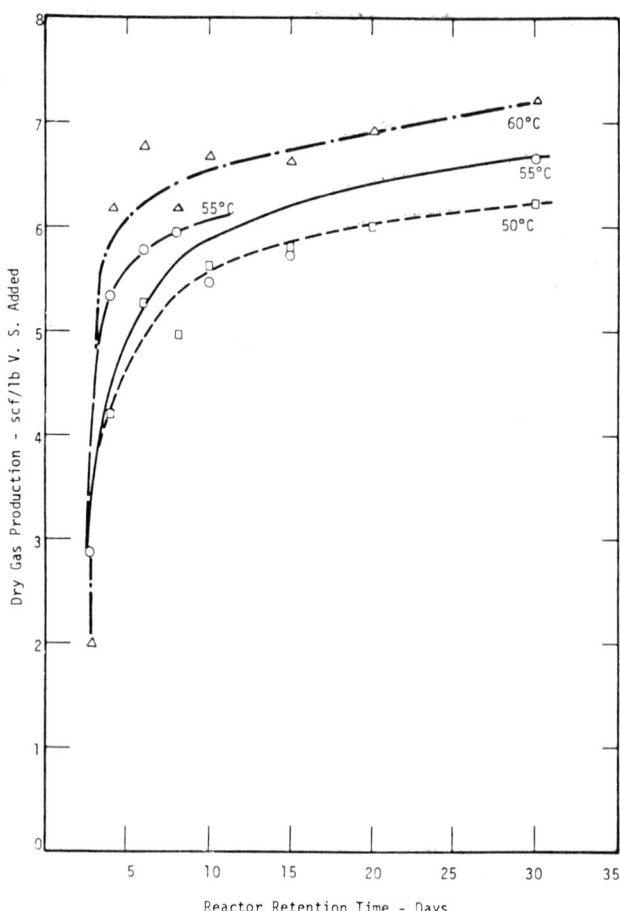

Source: PB 240 768

Digester Design

Tank Design: The digestion tank must be designed to provide the best possible growth conditions for methane-producing bacteria. The inner surface must be constructed of materials which will not be corroded by fermentation products or contribute toxic ions to the digestion mixture. Means must be provided for the introduction of feed and other additives, withdrawal of effluent, and collection of gas produced. Provision for withdrawal of culture samples during digestion would also be desirable. It must be possible to monitor pH values and operating temperature in the digester and to adjust those values during the digestion process.

Mode of Operation: Additions to and withdrawal from the digester can be made on a batch or continuous basis. Intermittent additions could be made daily or so frequently as to be considered a semicontinuous process. A batch process would be a constant volume process with reactants enclosed in the tank with no additions or withdrawals of volatile solids or waste products during the batch retention period. The headspace would be fitted with a valve permitting gases to flow out of the reactor as they are evolved so that a constant headspace pressure would be maintained.

In a batch process, the digester would be emptied at the end of the retention period. The batch process does have an advantage where the nature of the feed (solid waste) varies in physical or chemical properties or changes its properties during digestion and mixing. Each batch can be processed for a different amount of time or under different conditions, at the discretion of the operator, until digestion is satisfactory (20). Such a process would require a highly efficient bacterial culturing facility to shorten the attainment of a maximum stable state gas production level.

In a continuous digestion process, a feed and additives would be added continuously to the digester and a portion of the mixture continuously withdrawn. The anaerobic solid waste digestion process would probably be carried out on a semicontinuous basis, with feed added after an equivalent volume of reactor contents had been withdrawn.

Christopher (5) operated an anaerobic digestion process using simulated solid waste. He maintained stable digester operation with additions of small increments at frequent intervals. He found that the optimum length of time between additions was a function of particle size. Single large daily increments of finely divided solids degraded digester performance. With larger pieces of solid waste, longer periods between feed additions are possible.

Christopher believes that because of the greater surface area, bacterial attack on the finely divided solids resulted in production of acid at a rate greater than the ability of the symbiotic methane-producing bacteria to convert the acids to methane. With larger pieces of solid waste, the initial rate of acid production is lower because less surface area is exposed to bacterial attack. Because of the slower rate of bacterial degradation, an adequate supply of feed is available over a longer period and additions need not be as frequent.

Continuous operation is less flexible than batch since all material passing through the system receives the same treatment regardless of properties. Moreover, it may be difficult for the front-end to provide feed to the digester on a continuous basis.

With continuous operation, the digestion tank could be smaller. These potential savings may be outweighed, however, by the requirements of the continuous process for most costly equipment such as valves, piping and metering controls for introducing fresh feed and additives and for withdrawing digested material. A semicontinuous process, using intermittent additions and withdrawals would have many of the advantages of continuous operation while maintaining some improved level of flexibility. Such a process is considered to be most appropriate.

Antifoam Agents: Intermittent additions of antifoam agent might be necessary.

The bubbles of gas produced in the mixture are likely to form a stable foam if proteins are present in the digestion mixture or if a scum is allowed to form on the surface. If foam is not controlled, it will rise in the headspace and will be forced from the tank with the gas.

Two types of antifoam agents are available: inert (silicone compounds) and organic (oils and long chain alcohols). The inert type is preferable, since it may be added at the beginning of the digestion process, but inert antifoam agents are considerably more expensive. Organic antifoam agents are added only when foaming occurs and in amounts just sufficient to break the foam, since an excess of antifoam agent may be toxic to microorganisms.

Possible Modifications: The liquid effluent (filtrate or centrate) can be recycled into the digestion system for use as makeup water. In addition to saving water costs, recycling would reduce the volume of liquid effluent requiring further treatment and disposal. Heat can also be conserved by effluent water recycle. Additionally, as liquid effluent is returned to the digester, it would supply beneficial microorganisms. There is a danger, however, of a buildup of toxic materials.

Two-stage digestion is also a possibility. The first digester could optimize conditions for hydrolysis of organic compounds to volatile acids by acid-forming bacteria. The second digester could optimize conditions for the methane-forming bacteria. This would protect the sensitive methane producers from sharp changes in digester conditions due to variations in incoming waste.

Wolf (14) developed such a system on a laboratory scale. For the acid-forming stage, a conventional digester was overloaded with solids (0.52 pound volatile solids per cubic feet per day) and maintained at pH 6.0. Gas production stabilized at 1 cubic foot of gas per pound volatile solids or approximately 11% of the gas generation observed in a conventional digester. Composition of this gas was 98% CO_2. The volatile acid concentration was 10,000 ppm and included formic, acetic, propionic, isobutyric, n-butyric, isovaleric and n-valeric acids.

This effluent and a conventional thermophilic culture were fed to a second-stage digester which produced 10 cubic feet of gas per pound of volatile solids. The composition of the gas was 70 to 75% methane. With two-stage digestion, methane productivity was 7.4 to 7.7 cubic feet of methane per pound volatile solids. In a conventional single-stage digester, Wise (13) was able to attain methane production of only 5.3 to 4.7 cubic feet of methane per pound volatile solids.

SLUDGE DEWATERING

Methods

The most common methods of sludge dewatering are air-drying, vacuum filtration, or centrifugation. Air-drying is usually carried out in sand beds and the liquid is removed by two mechanisms: filtration through sand, and evaporation of the water to equilibrium moisture content. Air-drying of sewage sludge reduces the moisture content to about 65%. This method requires substantial land area and does not permit recycling of liquid effluent.

Christopher (5) found that air-drying of sludge on a sand bed and subsequent removal and disposal of the dried sludge as landfill was the most economical system for handling digester sludge.

In vacuum filtration, solids are separated from the liquid by means of a porous medium such as cloth, steel mesh or tightly wound coil springs which retain the solids. Variables affecting filtration rate and final moisture in vacuum-filtered sludge are:

- Pressure drop across sludge and filter
- Area of filtering surface
- Initial solids concentration
- Viscosity of filtrate
- Size and shape of solid particles
- Chemical composition of sludge and liquid
- Chemical conditioners, such as organic polymers which improve filterability

In centrifugation, the effect of solids concentration must be considered. Not every centrifuge machine is appropriate for dewatering slurries with high solids concentration. The mechanism of cake discharge must also be considered. During the experimental period, cake moisture and solids capture should be determined at various rotational speeds.

Supernatant

Since conditions in the digester facilitate microbial growth, the effluent will have a high level of contamination as evidenced by a high BOD or COD, and a high concentration of colloidal or suspended solids.

Suspended solids could be removed by a chemical coagulant. Colloids can be destabilized by chemicals such as alum or ferric chloride. Dissolved salts could be removed by reverse osmosis. Ultrafiltration with a high molecular weight cutoff membrane would allow passage of the dissolved salts, but the pore size would be sufficiently small to retain suspended and colloidal particles resulting in a water product with very low turbidity. The supernatant could be recycled into the digestion tank, or piped or transported to a sewage plant.

Cake

There is substantial potential for both ground and surface water pollution by the leachate from digester solids used in landfill. Cake used for landfill should be assayed for pathogenic microorganisms.

Not all the organic material in the feed will have been converted to methane and CO_2. The solids remaining after digestion and dewatering will contain a substantial quantity of energy. An energy balance should be run to determine the potential of sufficiently drying the cake to be able to use some of its fuel value to heat the digester. If the solids are incinerated, the residue or ash may be disposed of in landfill.

GAS PREPARATION

Desired Gas Characteristics

In order for the gas produced by anaerobic digestion to be acceptable by a gas utility customer, it must contain at least 90% methane and have a heating value of at least 1,000 Btu per standard cubic foot. Additionally, the gas must have an extremely low moisture content.

As discussed previously, CO_2 and CH_4 are theoretically produced in equal amounts during anaerobic digestion; however, the CO_2 has a much higher solubility potential than the CH_4. The quantity of CO_2 which remains in solution may be controlled by temperature of the digester, retention time, bicarbonate concentration, and pH. These parameters can be optimized within limits discussed previously, to reduce the CO_2 concentration in the resultant gas.

Based on laboratory studies, gas compositions such as those shown in Table 4.11 (8) were found to be possible. CO_2 content ranged from as low as 38% to a high of 70%.

Similarly, the moisture content of the digester gas will vary, principally as a function of temperature, from about 11% maximum by weight for thermophilic processes at 55°C to 5% maximum for mesophilic processes at 40°C.

Thus, if pipeline quality gas is to be produced by this UTM facility the raw digester gas will have to be scrubbed to eliminate both CO_2 and water vapor. Scrubbing costs can be high, both economically and in energy consumption. Dynatech estimated this cost for a 1,000-ton per day facility to be on the order of $2.3 million in capital costs (about 18% of the total plant costs) plus $270,000 per year in operating costs (about 4% of the total annual operating cost). Costs cited are 1974 dollars.

Energy required for gas scrubbing was estimated to be 17 million Btu per hour (about 54% of the total energy required to operate the facility or about 12% of the energy produced in the same time period) (21).

Carbon dioxide and water vapor scrubber technology is well understood today and mathematical models may be easily constructed to simulate the effects of varying raw gas characteristics on scrubber design, cost and energy requirements for a given level of removal of these substances.

Once the relationships between digester parameter variations, raw gas composition, total methane production, and cost have been determined, total system optimization may be undertaken. This later analysis may also take into account the possibilities of producing low-Btu gas for special application.

Gas Scrubbing Technology

In order to remove the carbon dioxide from the digester gas, wet or dry process scrubbers can be used. Two widely used methods of purifying natural gas are a bubble column [Girbotol Process (19)] and molecular sieve adsorbents. Gas purification methods are fully described by Kohl and Riesenfeld (23).

TABLE 4.11: REACTOR AND GAS VARIABLES

Reactor	Retention Time	pH	Alkalinity (mg/l CaCO$_3$)	Caustic Added (meq/day)	Percent Output CH$_4$	Percent Output CO$_2$	scf Gas/lb Volatile Solids Added
Reactor and Gas Variables at 60°C							
1	3	6.8	1,030	9.14	62	38	1.96
2	4	6.8	1,140	7.85	58	42	6.17
3	6	6.8	1,530	6.43	57	43	6.75
4	8	6.9	1,600	5.57	57	43	6.16
5	10	6.9	1,920	3.00	57	43	6.64
6	15	6.8	1,900	1.86	56	44	6.61
7	20	6.8	1,900	0.86	55	45	6.90
8	30	6.8	2,030	0.29	55	45	7.20
Reactor and Gas Variables at 50°C							
1	3	6.75	1,156	92	57	43	–
2	4	6.78	1,154	87	60	40	4.20
3	6	6.83	1,251	54.7	61	39	5.25
4	8	6.81	1,394	52.5	54	46	4.98
5	10	6.82	1,544	45.0	50	50	5.60
6	15	6.85	1,778	22.5	46	54	5.75
7	20	6.86	2,082	15.0	40	60	5.90
8	30	6.89	2,343	12.5	35	65	6.20
Reactor and Gas Variables at 40°C							
1	4	6.77	1,244	80.3	63	37	3.6
2	4	6.77	1,319	80.3	62	38	3.6
3	6	6.78	1,595	70.3	53	47	4.16
4	8	6.78	1,780	52.0	48	52	4.3
5	10	6.78	1,982	49.0	42	58	4.6
6	15	6.80	2,356	36.0	35	65	5.0
7	20	6.83	2,554	23.0	35	65	5.3
8	30	6.87	3,087	19.0	30	70	5.6
Reactor and Gas Variables at 35°C							
1	4	6.77	1,400	93	69.7	30.3	2.02
2	4	6.78	1,460	93	69.8	30.2	2.02
3	6	6.78	1,700	59	64.3	35.7	2.91
4	8	6.75	1,840	56	58.6	41.4	3.36
5	10	6.75	1,900	56	57.2	42.8	3.84
6	15	6.77	2,300	38	53.8	46.2	4.23
7	20	6.81	2,600	38	53.4	53.4	4.28
8	30	6.87	3,000	38	53.8	46.2	4.39

Source: PB 240 768

In the bubble column process (Figure 4.9) (21), carbon dioxide and hydrogen sulfide, if present, are absorbed in the aqueous low-temperature solution of an ethanolamine compound. This solution then enters a stripping column where the CO_2 and H_2S are released upon heating. The ethanolamine compound is then returned to the bubble column to absorb more acid gases. Three ethanolamine compounds are used in this process: monoethanolamine (MEA), diethanolamine (DEA) and triethanolamine (TEA).

FIGURE 4.9: GIRBOTOL PROCESS

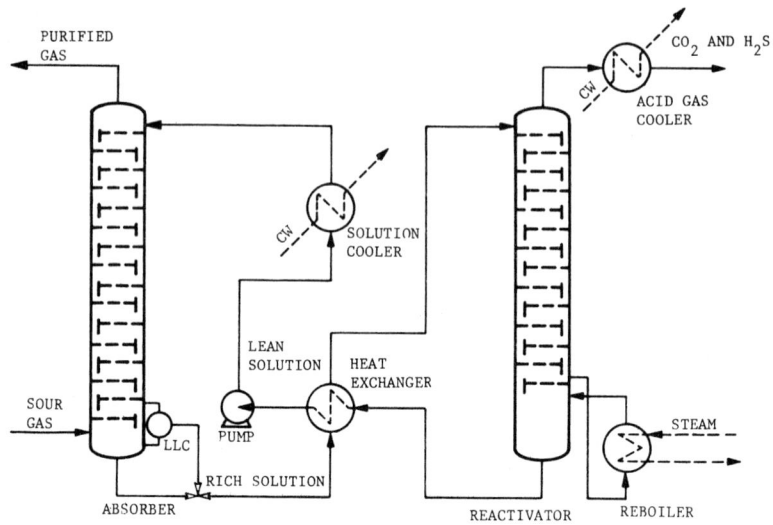

BASIC FLOW DIAGRAM

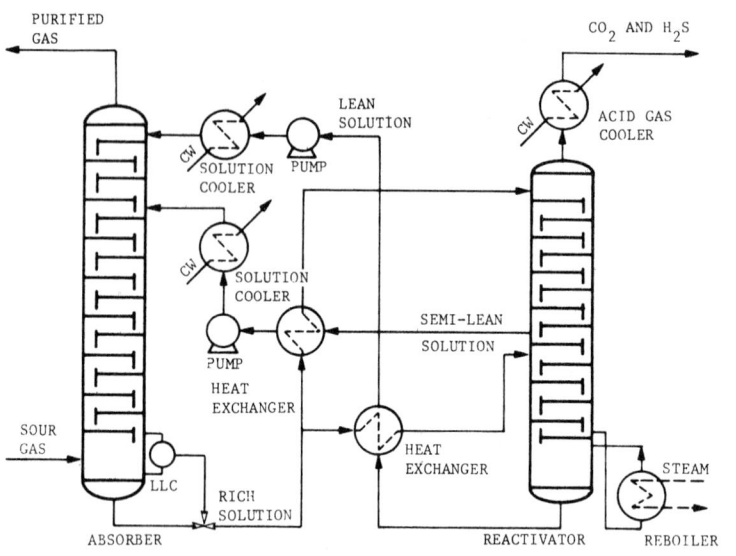

MODIFIED GIRBOTOL PROCESS (USING SPLIT STREAMS TO REDUCE STEAM CONSUMPTION)

Source: PB 240 768

These compounds have differing absorptive capacities for CO_2 depending upon their concentration in solution, the temperature of the solution, and the partial pressure of the CO_2 in the raw gas. A 15 to 20% aqueous solution of MEA was found to be the most suitable for processing natural gas. Compared to the other ethanolamines, MEA has a higher capacity per unit weight of solvent (although the difference is not great), a higher reactivity, and is more easily regenerated.

MEA also had disadvantages which set it apart from the other ethanolamines, particularly in the presence of carbonyl sulfide, where a heat-stable compound, diethanolurea, is formed, resulting in loss of amine. This reaction, however, should not be a problem in the anaerobic digestion process which produces no (or trace amounts of) carbonyl sulfide.

In addition to ethanolamines, sodium carbonate, potassium carbonate and potassium phosphate have been used in natural gas scrubbing facilities. These processes have generally been applied more to H_2S scrubbing than to CO_2 scrubbing; however, with the differing requirements inherent in the high CO_2 and water vapor characteristics of the raw digester gas, these technologies should also be considered.

Water vapor may be removed from the raw digester gas either before or in conjunction with the CO_2 scrubbing process. Depending upon the technology used, preremoval may raise the temperature of the raw gas, thus increasing the energy required for CO_2 removal. On the other hand, postabsorption drying will tend to cause dilution of the CO_2 scrubber working fluid as water is condensed out of the gas during this process.

Another gas purification method which allows for simultaneous CO_2 removal and drying is the use of molecular sieve adsorbents. In this process, polar adsorbents, such as synthetic zeolites (sodium or calcium aluminosilicates), ammonia, acetylene, hydrogen sulfide and sulfur dioxide are used. After adsorption, the zeolite is heated and the polar molecules are driven off. CO_2 content has been reduced to as low as 5 ppm through the use of this method.

Figure 4.10 (21) shows the adsorptive capacity of artificial zeolite for carbon dioxide at various partial pressures. Figure 4.11 (21) shows the adsorptive capacity of artificial adsorbents for water vapor at 10 mm partial pressure over a range of temperatures.

The plant required for this use of a molecular sieve adsorbent would have two or more containers so that while one container is adsorbing the water vapor and CO_2, the other may be heated to regenerate its contents.

As each container reaches its saturation point, it can be taken out of stream and heated for regeneration while one of the other vessels can be put on line in an adsorption mode.

The selection of the proper scrubber technology will depend on all of the above considerations as well as the digester operating characteristics and the energy required to regenerate the scrubber working compound.

FIGURE 4.10: ADSORPTIVE CAPACITY OF ARTIFICIAL ZEOLITE FOR CARBON DIOXIDE

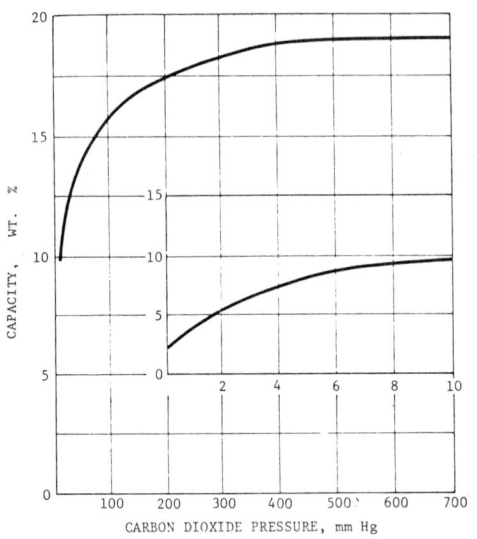

FIGURE 4.11: ISOBARS FOR SILICA, ALUMINA AND ARTIFICIAL ZEOLITE (MOLECULAR SIEVE) TYPE ADSORBENTS

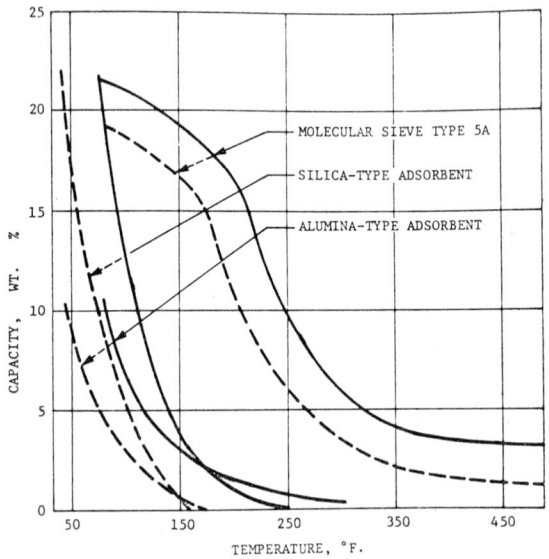

Source: PB 240 768

REFERENCES

(1) Boettcher, R.A., *Air Classification of Solid Wastes,* USEPA, Office of Solid Waste Management, 1972, p 33.
(2) Ibid, p 49.
(3) Drobny, N.L., et al, *Recovery and Utilization of Municipal Solid Wastes,* USEPA, Office of Solid Waste Management, 1971, p 26.
(4) National Center for Energy Management and Power, University of Pennsylvania, "Technology for the Conversion of Solar Energy to Fuel Gas," January 1973.
(5) Christopher, G.L.M., "Biological Production of Methane from Organic Materials," final report to Columbia Gas Service System Corporation, 1971.
(6) Pfeffer, J.T. and Liebman, J.C., "Biological Conversion of Organic Refuse to Methane," Report No. NSF/RANN/SE/G1-39191/PR74/2, July 1974.
(7) Cooney, C.L., "Thermophilic Anaerobic Digestion of Solid Waste," Bioconversion Energy Research Conference at University of Massachusetts, Amherst, Mass., June 25-26, 1973.
(8) Pfeffer, J.T., *Reclamation of Energy from Organic Refuse,* NSF/RANN, April, 1973.
(9) Casida, L.E., *Industrial Microbiology,* New York: John Wiley and Sons, Inc., 1968.
(10) McCarty, P.L., "Anaerobic Waste Treatment Fundamentals. Part 2—Environmental Requirements and Control," *Public Works,* 95:123-126, October 1964.
(11) McCarty, P.L., "Anaerobic Waste Treatment Fundamentals. Part 3—Toxic Materials and Their Control," *Public Works,* 95:91-98, November 1964.
(12) Golueke, C.G., "Temperature Effects on Anaerobic Digestion of Raw Sewage Sludge," *Sewage and Industrial Wastes,* Volume 12, No. 25, 1958.
(13) Wise, D.L., "Fuel Gas Production from Solid Waste," Bioconversion Energy Research Conference at University of Massachusetts, Amherst, Mass., June 25-26, 1973.
(14) Wolf, M. and Keenan, R., "Conversion of Solar Energy to Fuel Gas," Bioconversion Energy Research Conference at University of Massachusetts, Amherst, Mass., June 25-26, 1973.
(15) Eckenfelder, W.W., "Mechanisms of Sludge Digestion," *Water and Sewage Works,* Volumn 114, No. 6, pp 207-210, June 1967.
(16) McCarty, P.L., Kugelman, I.J. and Laurence, A.W., "Ion Effects in Anaerobic Digestion," Department of Civil Engineering, Stanford University, March 1964.
(17) Schulze, K.L., "Studies on Sludge Digestion at Increased Solids Concentrations," *Sewage and Industrial Wastes,* Volume 30, No. 1, 1958.
(18) Eckenfelder, W.W. and O'Connor, B.J., *Biological Waste Treatment,* Pergamon Press, New York, 1961.
(19) McCarty, P.L., "Anaerobic Waste Treatment Fundamentals. Part 4—Process Design," *Public Works,* 95:95-99, December 1964.
(20) McCabe, W.L. and Smith, J.C., *Unit Operation of Chemical Engineering,* New York: McGraw-Hill Book Co., Inc., 1956.
(21) Kispert, R.G., et al, *Fuel Gas Production from Solid Waste,* Dynatech R/D Co., Cambridge, Mass., July 1964.
(22) Strauss, W., *Industrial Gas Cleaning,* New York: Pergamon Press, 1966, p 97.
(23) Kohl, A.L. and Riesenfeld, F.C., *Gas Purification,* New York: McGraw-Hill, 1960.

ANIMAL WASTE DIGESTION

The sources of the material in this chapter are the following reports: PB 240 113, PB 231 149 and PB 240 768. For a complete bibliography of these reports see page 222.

OREGON STATE UNIVERSITY CONVERSION SYSTEM BASED ON THERMAL DISCHARGES

Animal Waste Problems

A major problem confronting animal agriculture is adequate disposal of waste. In the past, livestock were produced by many farmers feeding small numbers of animals. These dispersed operations kept manure concentrations relatively small. The manure was applied to cropland on the farm where it was produced. Changing practices have created situations where the production of animal waste occurs in large quantities on small areas.

This concentration is developing in the form of outside feedlots and total confinement rearing units where thousands of animals are contained continuously on a few acres of land. The capacity of the local environment to dilute, stabilize and dissipate the accumulation of wastes from these systems is exceeded in many locations.

There are about 110 million cattle, 22 million sheep and goats, 54 million swine, 825 million chickens and turkeys, and 7 million horses in the United States. These produce five million tons of waste daily, containing over 800,000 tons of dry matter, 32,000 tons of nitrogen, and 10,000 tons of phosphorus. In comparison, the 210 million humans in the United States produce about 375,000 tons of waste containing 59,000 tons of dry matter, 2,900 tons of nitrogen and 350 tons of phosphorus daily (1)(2).

Current manure management is expensive in terms of equipment and labor, and frequently does not provide adequate protection to the environment. Highly mechanized manure management systems which give adequate consideration to environmental quality are essential for the future of livestock production if an abundant supply of high quality meat is to be available at a reasonable cost.

Malodors emanating from animal production units are frequently disagreeable to nearby residents in both rural and urban communities. In several cases such malodors have resulted in legal procedures leading to injuctions to halt the operation of animal units unless some method of waste disposal is followed that controls the malodors.

Public action programs for pollution abatement make it necessary to develop methods for controlling odors and the concentration of solids, nitrogen and phosphorus which enter surface and ground water. Animal wastes are also a potential source of pathogenic organisms. These (organisms or wastes) can be scattered into the environment and become a health hazard, particularly under the increasing demand for water-based recreational sites.

The need for intensified research efforts in animal waste management was recognized about 10 or 12 years ago. Since then a series of conferences devoted to this topic have been held, indicating wide-spread concern. The urgency of the problem has been discussed, as well as the results of various studies on the characteristics, handling, treatment, utilization and disposal of animal manures. Animal manure is a highly concentrated source of organic matter, plant nutrients and microorganisms (3)(4). The quantitative character of waste produced by a specific livestock operation is a function of the number and species of animals being raised in addition to the system of manure collection and the ration being fed.

Runoff from unroofed cattle feedlots has been studied to determine quantity and quality as a function of feedlot design, management and climatic conditions (5). Similar information for other species of livestock being confined in unroofed pens has not been obtained. Systems for the collection of cattle feedlot runoff have been designed which minimize the quantity of runoff for disposal (6).

The storage and treatment of livestock manure are accompanied by odors which are objectionable to both livestock producers and the public (7). Some work has been done in characterizing these odors by identification of the component gases (8)(9). Measurement of odors to determine nuisance levels and progress in odor control have been of limited success (10)(11)(12).

Different methods of facilitating manure collection in livestock housing have been evaluated. Sloping floors, flushing gutters and various designs of slotted floors have been used in commercial installations. Similarly, manure storage tanks of various capacities, configurations, and structural materials have been constructed. Storage remains an expensive aspect of manure management, particularly in northern states where long-term storage is required.

Various devices for the agitation of stored manure and its transfer to cropland have been considered. Centrifugal and positive displacement pumps, augers and vacuum systems for transferring manure from storage tanks to hauling vehicles were studied (13). Tank wagons and irrigation pumps for the distribution of manure to cropland are under study (14). These systems are currently limited by climatic conditions, odor problems, large land area requirements, and cost. The escape of toxic gases during manure tank agitation has been of concern to many operators and has resulted in animal deaths (7).

Various schemes have been proposed for the treatment of animal manure prior to utilization or disposal. Lagoons were found to effectively store manure but did not produce an effluent suitable for discharge into streams. In addition, lagoons have been sources of objectionable odors (15)(16)(17)(18).

More sophisticated anaerobic treatment units have been considered but not adopted by the livestock industry (19). Aerobic treatment of animal manures has been of considerable interest during the past few years. Aerated lagoons and oxidation ditches are effective in minimizing odors from waste treatment.

However, they do not produce effluents suitable for discharge to surface watercourses and they have high rates of energy consumption (20)(21)(22)(23). Treated animal waste remains high in organic matter and plant nutrients. Land application is the most suitable disposal method for this liquid.

Research on the utilization of animal manure for purposes other than direct soil enrichment is under way. Two examples are the refeeding to livestock (24) and the use as a substrate for the growth of economically useful plants. Alternate methods of manure processing under consideration include dehydration and incineration (25).

These methods are in exploratory stages and must be pursued through the development of usable equipment and the planning of auxiliary equipment and management techniques. Research is continuing on the use of soil as a recipient for animal manures. More sophisticated methods of manure application, more effective use of the nutrients, and use of the soil as a high capacity treatment system are being investigated. (26)(27).

Animal waste management research has been expanding rapidly and an increasing number of papers are published annually. Reviews of these papers are published for easy access (1)(2). Loehr (28) published a comprehensive review of animal waste research. Muehling (29) prepared a review of research on swine wastes.

Photosynthetic reclamation of agricultural solid and liquid wastes was investigated in a pilot scale project at the University of California (30). A partially closed system of animal waste management based on the integration of an anaerobic and aerobic phase, the recycling of water, and the reclamation of algae was studied at the Sanitary Engineering Research Laboratory. The investigators present a detailed materials balance which indicates a favorable economic outlook for the proposed system. A report of this study is given in the next section.

Proposed Animal Waste Management System

A study has been done by L. Boersma et al at the Water Resources Research Institute, Oregon State University, Corvallis, Oregon (PB 240 113) to determine the feasibility of using waste heat from steam electric plants as a low cost source of energy for maintaining stable, elevated temperatures for anaerobic digestion of animal wastes. As a result of the study, an animal waste management system was proposed which would produce both a methane-rich fuel gas and a high protein animal feed. The following description of the system is concerned primarily with its gas-producing function.

Design of the proposed waste management system is based on the concept of a livestock confinement building where the manure is quickly removed from the animal quarters. The manure will be hydraulically flushed from the building with a frequency selected to prevent anaerobic decomposition and the associated odors. The flushing water used in the system will be recycled through a treatment scheme as outlined in Figure 5.1. The animal unit should be designed so that the animals have no access to the manure transport water, eliminating potential disease problems associated with drinking partially treated wastewater.

The proposed system (Figure 5.1) consists of a manure handling unit, nutrient recovery unit, and soil filter unit. Manure is flushed from the animal quarters by the outflow from siphon-activated storage tanks, to a sump. It is then pumped with an air-lift pump to a solid-liquid separator. The solid slurry falls into an anaerobic digester and the liquid flows into algal culture basins after filter treatment to decrease turbidity and remove color caused by dissolved organic matter.

Water from the algal culture basins is passed through a centrifuge for separation from the algal mass, then stored in a reservoir. From there it can be pumped to the flush tanks for reuse in the flushing process. Gases from the anaerobic digester are vented to a storage tank for later use or to a soil filter for the removal of odors. Excess water is applied to a soil filter field.

The proposed system of animal waste management will be analyzed by considering the use of swine manure produced at the Swine Research Center on the Oregon State University campus. The discussion will be based on the management of manure produced by 50 swine. This unit size was selected because it was considered large enough to produce problems similar to those of real-world operations.

Manure Handling System: Animal Units — The conceptual design of the proposed system of waste management will be developed by considering the management of swine manure from a portion of the confinement finishing building at the Oregon State University Swine Research Center, 2.5 km west of the Corvallis campus.

Hogs are kept in this building on partially slotted floors. The manure drops into 1.20 meter wide pits from which it is periodically drained to a nearby anaerobic lagoon. Pens, 3 meters wide and 2.15 meters deep, are along a central 1.20 meter wide walkway. Automatic waterers and self-feeders are in each pen. These pens would be modified by replacing the pit floor. The new floor would slope at a rate of 2%, allowing the pens to be flushed using flush tanks (Figure 5.2).

Flushing should occur every two to four hours, depending on the odor control achieved within the building. Effluent from the nutrient recovery system is brought into the flush tanks with a timeclock-controlled pump. A siphon is engaged when the flush tanks fill to a prescribed level, and the tank contents are discharged into the gutters with sufficient velocity to carry manure out to a sump (Figure 5.3).

FIGURE 5.1: SCHEMATIC DIAGRAM OF THE PROPOSED ANIMAL WASTE MANAGEMENT FACILITY

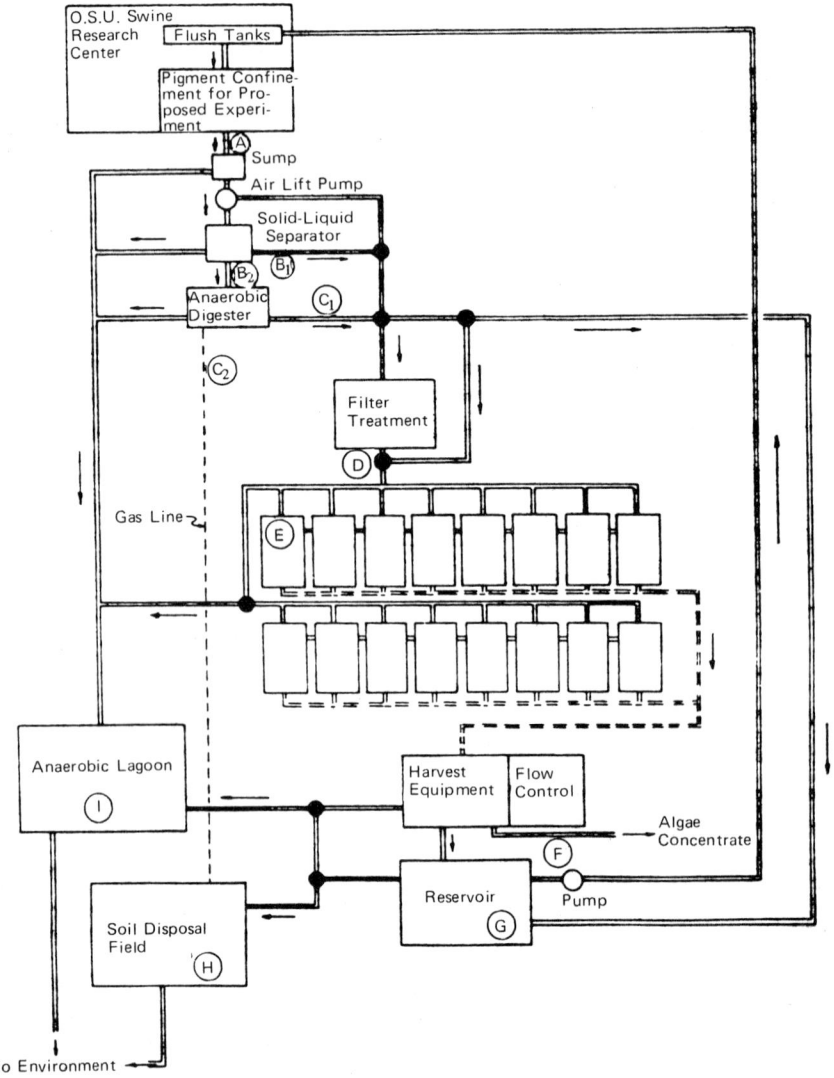

Source: PB 240 113

The sump functions as a temporary storage reservoir for the manure after flushing. This method of manure removal has been successfully used previously and is adaptable to many buildings.

Animal Waste Digestion   143

FIGURE 5.2: CONCEPTUAL DESIGN OF THE SIPHONS USED TO ACTIVATE
THE MANURE FLUSH TANKS

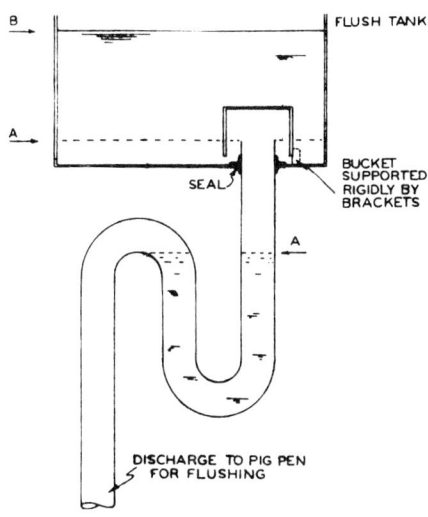

FIGURE 5.3: SCHEMATIC OF THE AIR-LIFT PUMP ARRANGEMENT USED
TO LIFT THE MANURE TO THE SOLID-LIQUID SEPARATOR

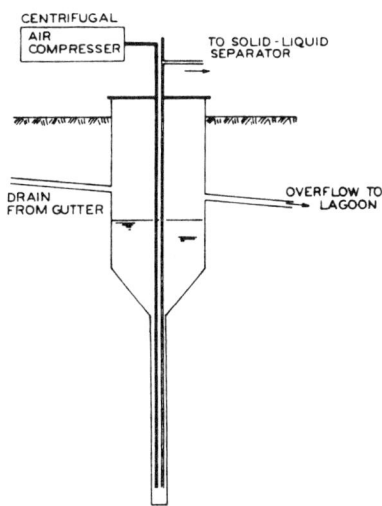

Source: PB 240 113

Manure Input — Six pens of the design described above would have a combined capacity of 50 animals with an average weight of 55 kg each. The anticipated rate of manure production is shown in Table 5.1. The manure will be flushed from the pens to a sump, then pumped with an air-lift pump to a solid-liquid separator located over an anaerobic digestion tank (Figure 5.3). The air-lift pump was selected to achieve a high degree of reliability and to lift the slurry at a uniform rate of 100 liters per minute or less, depending upon the flushing frequency desired.

TABLE 5.1: DAILY RATE OF MANURE PRODUCTION IN THE TEST FACILITY

Parameter	Value
No. of swine	50
Average live weight	55 kg
Total live weight	2,750 kg
Manure	136 kg dry weight
Solids content	15%
Total solids	20.4 kg dry weight
Volatile solids	17.0 kg dry weight
BOD	5.7 kg
Nitrogen (as N)	1,134 grams
Phosphorus (as P)	227 grams
Potassium (as K)	136 grams

Source: PB 240 113

Solid-Liquid Separator — Untreated swine manure contains a large quantity of particulate matter which is resistant to breakdown under aerobic conditions. If not removed prior to introduction into the algal basins, this material would increase turbidity, settle out as a sludge layer, and finally result in a filler of low nutritional value in the harvested algae. The purpose of the solid-liquid separator is to remove these particles and direct them to the anaerobic digester where they can be liquefied more efficiently. The separator concentrates the solids to a 5% slurry which is suitable for feeding the digester.

The solid-liquid separator selected for use in the proposed management system was recently developed at Oregon State University for the removal of coarse solids from animal manures (31). It is intended for use in water recycling systems where extensive particulate matter that could interfere with the functioning of conventional water handling equipment is troublesome.

The device consists of a 70 centimeter diameter cylinder with flighting 15 centimeters deep attached perpendicularly to the inside surface of the cylinder (Figure 5.4). The cut-out portion of the cylinder shows the flighting, which has a continuous winding from the lower end to the upper end of the cylinder).

In operating position, the cylinder is supported at an incline of approximately 17° and rotated at the rate of 0.2 to 0.5 rpm. The design of the flighting and the rate of rotation are chosen to allow solids to settle into the spaces between the vanes while the liquid passes over the upper edges and flows out at the lower

Animal Waste Digestion 145

end of the device. The settled solids are worked upwardly by the rotation of the cylinder and dropped from the upper end in the form of a slurry into the inlet box of the anaerobic digester.

FIGURE 5.4: SCHEMATIC DIAGRAM OF THE ROTATING FLIGHTED SOLID-LIQUID SEPARATOR

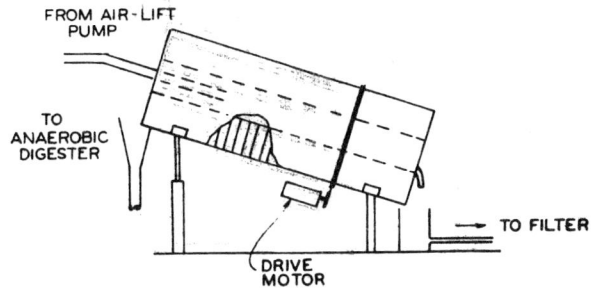

Source: PB 240 113

Anaerobic Digestion Tank — The slurry containing the solid particles drops into the inlet box of an anaerobic digester (Figure 5.5) whose purpose is to convert the solids to a liquid from which nutrient recovery is possible.

FIGURE 5.5: SCHEMATIC DIAGRAM OF THE ANAEROBIC DIGESTER

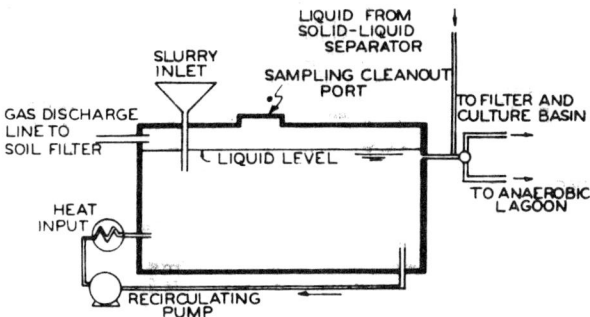

Source: PB 240 113

A water seal will be provided to prevent the escape of gases through this port. The contents of the digestion tank need to be continuously mixed and maintained at a temperature of 35° to 37°C. Both heating and mixing can be accomplished by passing a recirculating pump discharge through an exterior heat exchanger, using power plant waste heat or other energy sources. The effluent from the anaerobic digester will be subsequently added to the effluent from the solid-liquid separator.

The anaerobic digester in this report was designed on the basis of a daily loading rate of 1.00 kg of volatile solids per cubic meter. The conservative rate minimizes the potential for digestive upset. The volume required to accept the 17.0 kg of volatile solids produced daily is 17 cubic meters. Thus, a tank 3.0 x 4.0 meters in plan, 2.0 meters deep is specified. The water depth will be about 1.5 meters. The tank will be covered. Gases produced can be stored for later use.

Gas Storage — The design discussed here is based on the rate of manure production of 50 pigs. The rate of gas production from the anaerobic digester would be small and provisions for storing the gas for later use were not made. An energy balance of the proposed system should include an analysis of the use which can be made of the digester gases as an energy source in the system. The gases would probably best be used as an energy source fo the algal harvesting device.

System Safeguards — Precautions must be taken to insure manure removal when breakdown occurs in one of the various units of the system. System safeguards in the form of an anaerobic lagoon seem appropriate. Such a lagoon exits at the site selected for the demonstration of this concept. Overflow lines receiving raw waste, will be constructed at the sump at the control box ahead of the algal culture basins, and at the water storage tank. These lines will be designed to discharge directly into the anaerobic lagoon when diversion is necessary.

Flushing Gutter Malfunction — The manure handling system to be used was selected for its simplicity and freedom from normal operating difficulties. In spite of this, difficulties may be anticipated. Among the most likely are plugging of the sewer line or flush tank malfunction.

The drain line should be designed and constructed for easy cleaning and repair. The dosing siphon historically has been a trouble-free device once properly installed and adjusted. Regular cleaning and biological growth removal with normal cleaning solutions will help assure its continued function.

Failure of the Air-Lift Pump — Any pump designed to lift manure slurry may be expected to malfunction. The air-lift pump was selected because it is reliable and easy to repair when malfunction does occur. No moving parts are in contact with the manure slurry, thus minimizing accelerated corrosion and clogging.

Hog hair, straw, sticks and other foreign material may collect across the inlet to the pump. When this occurs, it will be possible to lift the entire pump assembly from the pit for cleaning. In the pump chamber is an overflow line directing excess water to the anaerobic lagoon if failure occurs during the night.

Anaerobic Digester Failure — Among the most common failures of waste handling systems has been anaerobic digester upset. The digester design presented above is conservative to reduce the likelihood of digester upset. The temperature of the anaerobic digester should be maintained at the optimum 35° to 37°C level. Minimal sludge accumulation within the anaerobic digester is anticipated. When this does occur the tank will be accessible to pumping with a diaphragm pump. When necessary the sludge can be transferred into the anaerobic lagoon to prevent excess accumulation.

BIOCONVERSION STUDIES AT THE UNIVERSITY OF CALIFORNIA (BERKELEY)

Work done at the Sanitary Engineering Research Laboratory (SERL) of the University of California (Berkeley) on the production of methane as a phenomenon of anaerobic digestion dates from the mid-nineteen-fifties and has continued with a few interruptions. The results of the work have been reported in a number of publications (32)-(41). The studies conducted at SERL can be divided into three sections, namely: digestion of animal wastes, digestion of domestic refuse, and digestion of sewage-grown algae.

The following material on the digestion of animal wastes is taken from a report made by C.G. Golucke of SERL at the Bioconversion Energy Research Conference held at the University of Massachusetts, Amherst, on June 25-26, 1973 (PB 237 149).

The concept of digesting animal wastes to produce a combustible gas, namely methane, is certainly not a new one. During World War II and for a short time thereafter, a number of small-scale digesters constructed to produce gas for kitchen use could be found in Germany and France (42)(43).

More recently, the idea has been revived by investigators in India, namely Singh (44) and Patel and Patel (45). Their papers present detailed designs for the construction of digesters and gas collectors scaled to use on small farms or groups of two or three farms.

According to Singh, one could expect about 1 cubic foot of gas to be produced per pound of fresh cow manure introduced into the digester. Patel and Patel cite a figure of 1.5 to 2 cubic feet per pound of fresh chicken manure. In both cases the methane content of the gas would be in the order of 65 to 70%.

The use of anaerobic digestion other than in lagoons as an animal waste treatment device has not found much favor in the United States. This observation is confirmed by the dearth of papers presented on the subject during the International Symposium on Livestock Wastes in 1971 (46); and by the unfavorable assessment made by Loehr (47).

Reasons for the absence of interest may be gleaned from Hart's paper on the digestion of livestock wastes (48). Chief among the reasons is the low order of the digestibility of wastes from livestock. The factor responsible for the limited digestibility of cattle--or any ruminant--wastes is quite obvious.

Readily digestible material is broken down in the animal's rumen, and hence its excreta would contain only that fraction of the animal's feed intake which is difficult to decompose. Other difficulties observed by researchers are mechanical in nature. For example, bedding material has a tendency to float and become entrapped in the scum layer. As a consequence, the scum layer becomes thick enough to clog gas ports and other outlets.

Investigations at the University of California, on the anaerobic digestion of animal manures covered two aspects, namely, the use of digestion as the sole disposal process, and digestion as a component of an integrated photosynthetic nutrient recovery system.

Digestion as a Sole Disposal Process

The incentive for studying the use of digestion as a means of treating animal wastes stemmed from an awareness of the urgency of the national need to develop a means of satisfactorily processing the wastes produced in confined animal-rearing operations. Of particular interest was the possibility of incorporating the treatment of such wastes along with municipal sewage solids in a common digester.

In the study, the digestion of chicken manure and that of steer manure were investigated individually and in combination with sewage sludge and solid waste components. When used in combination with solid waste components, the manures were intended to serve as a nitrogen source in the digestion of nitrogen-deficient wastes. The study was mainly on a laboratory scale.

Digester vessels used in the study and in those on the digestion of municipal wastes were four-liter glass vessels equipped for mechanical stirring and agitation of culture by gas recirculation. Gas was collected by water displacement and its composition was determined by gas chromatography. The experimental procedure was basically that followed in conventional digestion of raw sewage sludge. The detention period was 30 days.

The chicken manure had an average nitrogen content of 3.2%; carbon concentration of 23.4%; moisture content 9.8%; volatile solids 56.2%; and inorganic material 43.8%. (Unless otherwise specified, all concentrations and percentages cited are on a dry-weight basis).

The following materials and combinations of materials were used in the tests: 100% sludge; 100% steer manure; 50% steer manure, 50% grass; 50% steer manure, 50% chicken manure; 100% chicken manure; 50% sludge, 50% chicken manure; 50% chicken manure, 50% paper pulp; 31% chicken manure, 69% paper pulp; and 26% chicken manure, 74% paper pulp.

The average gas production by the digester receiving only steer manure was 1.4 cubic feet per pound of volatiles added. When chicken manure was digested the yield was 5.0 cubic feet per pound. The relative efficiency of the animals in utilizing the energy content of their diet was reflected by the difference in gas production from the digestion of their wastes.

Using as a basis gas production per pound of added volatile sewage solids in a digester fed sewage sludge, the digestion of steer manure was 14.9% efficient, and that of chicken manure 53.2% efficient. In terms of volatiles destroyed, the gas yield was higher from the digester fed chicken manure than that from the sewage sludge digester.

Although the manures were stabilized by digestion and rendered inoffensive, reduction in volume of the material was very slight. Destruction of total solids was less than 15%. The composition of the gas from the digesters fed manure was comparable to that of the gas from the sewage sludge digester. The methane content of the gas from the digestion of manure was only slightly less than that from the digestion of raw sewage sludge.

Digestion as a Component in a Photosynthetic Treatment System

The photosynthetic treatment system was devised to convert the nutrients in animal wastes into algal cells by way of bacterial activity and photosynthesis by the algal cells, and at the same time to recycle water used in maintaining sanitation in the animal's quarters.

The role of digestion was to stabilize the solids capable of settling in the washings from the manure troughs, and to generate methane, which could be burned to supply the heat needed to dry the algal product. A diagrammatic sketch of the pilot plant operated in the study is shown in Figure 5.6.

The pilot plant studies were preceded by a laboratory investigation of the digestibility of chicken manure, and a determination of those environmental conditions that would result in effective stabilization and maximum gas production. These determinations were made despite the fact that the anaerobic digestion of poultry wastes has been fairly well explored.

It was felt that variations in the reported results were of sufficient magnitude to warrant further investigations, at least to the extent of relating expected digestion characteristics to the photosynthetic treatment system.

Although design parameters also have been fairly well developed, those reported in the literature were not entirely applicable to the system. Applied loadings ranged from 0.03 to 0.120 pounds of volatile solids per cubic foot of culture volume per day. Fifteen-day and thirty-day detention periods were tried.

Gas yields in the laboratory experiments ranged from 4.5 to 9.21 cubic feet per pound of volatile solids introduced. An important observation made in the study was the need to digest the manure as soon as possible after it had been excreted by the animals (birds in this case).

Judging from the laboratory results with the chicken manure, from 25 to 33% of the volatile solids had been destroyed during the collection of the manure at the chicken farm and transportation to the laboratory. The results further showed that those volatile solids not destroyed during storage or drying were very resistant to further decomposition in anaerobic digestion.

Gas production in the 400-gallon pilot plant digester averaged almost 12 cubic feet per pound of volatile solids introduced. The methane content at the start of the pilot plant run was only 15% of the total gas produced. As the population of methane producers grew, the methane fraction increased until by the end of the run it was 55% of the total.

Judging from the laboratory results, the methane content undoubtedly would have reached 65 to 70% of the total. The high yield of gas per pound of volatile solids in the 400-gallon digester was quite probably a function of the manure solids introduced into the digester. The manure was transferred from the feeding troughs into the digester within an hour of its generation.

FIGURE 5.6: POULTRY OPERATION WITH AN INTEGRATED SANITATION WASTE MATERIALS RECYCLING SYSTEM

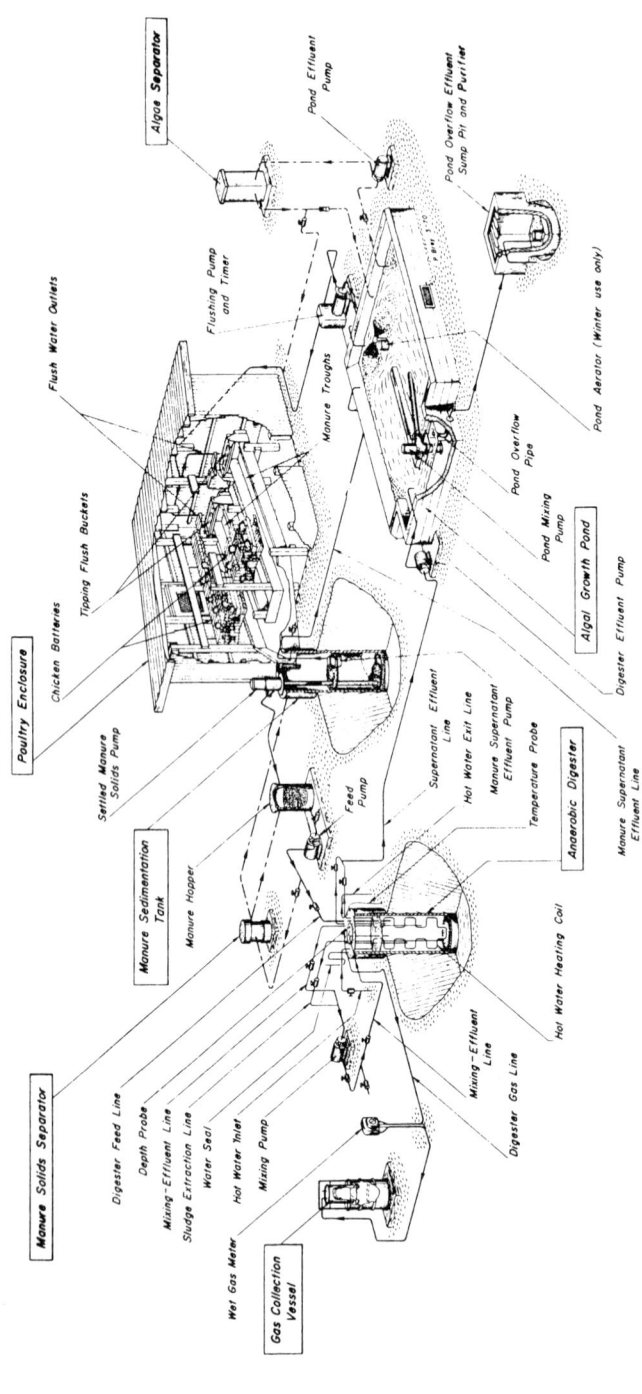

Source: PB 231 149

DEPARTMENT OF AGRICULTURE STUDY

G. Christopher of the United Aircraft Research Laboratory and M. Tuck of Hamilton Standard Corporation gave the following report on animal waste conversion at the Bioconversion Energy Research Conference of June 1973 (PB 231 149).

Hamilton Standard Corporation continuously operated two 20-liter anaerobic fermentors for a period of twelve months in a test program, sponsored by the Northern Regional Research Laboratories of the United States Department of Agriculture, Peoria, Illinois, utilizing cattle waste for the purpose of evaluating the suitability of this technology as a waste management technique.

The primary objectives of the program were: [1] to identify the significant parameters signalling imminent process failure and the proper corrective action to be applied with anticipation of incorporation of automatic process control; [2] to investigate high loading rates as an implied result of the increased reliability and stability of the process demonstrated during the early part of the program; and [3] to define an automatic control concept which would allow use of technology in agricultural-type situations.

Various "challenges" were imposed on the fermentors to study their effect upon the general fermentative process. These challenges were meant to be representative of actual operating conditions likely to be encountered during operation of a full sized facility. The challenges included: decrease of residence time by addition of both solids and water; increase of volatile solids loading; increase and decrease in operating temperatures; addition of minerals such as phosphorus; addition of both urea and urine.

Considering the widespread belief that anaerobic fermentation processes are inherently unreliable and unstable, one of the more important findings of this program was that continuous stable operation of the fermentors at loading levels as high as 16.0 grams volatile solids per liter per day was achievable. At no time during the imposition of these "challenges" were the fermentors in imminent danger of failure.

Besides the proven stability and reliability of fermentors operating at loading levels as high as 16.0 grams volatile solids per liter per day other pertinent results found were:

- [1] Gas rate (volume per mass of volatile solid) appeared to be the most sensitive of the analytical or measured parameters for recognizing fermentor "health" during operation.
- [2] Approximately 90 to 95% of both the carbon and total mass introduced was accounted for in the gaseous and output solids.
- [3] Approximately 90% of the nitrogen introduced was accounted for with no detectable loss as gas.
- [4] No significant amount of gaseous hydrogen was measured.
- [5] The gas generated consistently contained 52% methane and was producible at a demonstrated rate of 0.5 liter per gram volatile solids feed. In wastes that are 95% volatile, this results in an availability of about 1,770 calories per gram of wastes.

The one significant conclusion reached was that this process is a stable and reliable method of treating animal waste which can be used in an agricultural environment without the need for complicated controls or highly trained personnel.

OTHER REPORTS ON ANIMAL WASTE DIGESTERS USED FOR ENERGY

In two articles in *Compost Science* (44)(49), R.B. Singh describes a biogas plant for generating methane from animal wastes. According to Singh, a methane-producing gas plant is a relatively simple and inexpensive source of energy for farms since it will convert dung and vegetable waste into cooking and heating gas.

Once the desired methane production capacity is decided upon by user needs, the choice of type of plant can be made. Different systems can accommodate approximately 500 cubic feet of gas per day and are available with one or two chambers. Dung and vegetable matter should be mixed so that the carbon to nitrogen ratio is not more than 30:1.

Dry weight of dung should always account for at least half the total. The slurry should be agitated if vegetable waste is present, and should be kept warm enough for fermentation. Especially in cold climates or in winter months, artificial heating may be needed. The single chamber model should be emptied before gas production is quite complete, thereby preventing solids from settling out.

The fertilizer by-product of methane production should be liquid. In a double-chamber digester, slurry from which most of the gas has been released is passed to a second tank, where no agitation occurs. The second tank permits the solids to settle out. Gas is collected in small plants and 2-stage large plants by a metal drum inverted over the slurry. As the pressure builds up, the drum rises until the valve releases the gas and then settles again.

Five biogas plants are presented: the single-stage, double-chamber, small-scale biogas plant for producing about 100 to 500 cubic feet of gas per day; the two-stage digester which should be built only if dung is available from at least 50 cows; the large single-stage plant, which provides the thoroughness of the two-stage plant in the compactness of a single digester; the single-stage multiple-digester plant which is more efficient at producing gas and not designed for 100% digestion of dung for fertilizer; and the vegetable-digesting gas plant, the most convenient because of its small size and simple construction.

Two or three digesters should be installed, because these are not continuously producing plants. All five plants are described in detail, recommendations are made for their construction, and most efficient and productive utilization.

Due to the serious environment problems associated with the large concentrations of manure produced by poultry raised in confinement, the idea of being able to use chicken manure to produce methane and use it as a source of fuel is attractive.

In a paper appearing in the *Journal of the Water Pollution Control Federation* (50) C.W. Savery and D.C. Cruzon discuss the design of an anaerobic chicken manure digester and the experimental method used to produce methane from chicken manure. The result indicates that 130 liters of gas, 69% of which was methane, was produced per kilogram of wet manure. It is concluded that due to strict pollution controls on poultry farms, it is economically feasible for a poultry farm to invest in an anaerobic digester to produce methane to provide energy for the farm and to control pollution.

An article in *Pennsylvania Econotes* (51) discusses how Lebanon County in Pennsylvania is producing methane gas from animal manure. An experimental anaerobic digester has been built to demonstrate that process. The demonstration model was constructed for these reasons: to demonstrate the energy-producing potential of agricultural wastes; to encourage interest in using anaerobic digesters as part of a farm's system for holding and disposing of manure; and to show that the use of methane fuel can alleviate air pollution.

The production and use of methane from sewage sludge and animal wastes has been a common practice in Europe since 1930. The equipment used for this study is described in detail. In addition to the fuel benefits derived from the use of methane gas, research data indicate that the residue which remains after gas production has the same nitrogen content as does the raw waste, thereby retaining its value as a crop fertilizer. The economics of this conversion are also outlined in the report.

T.H. Jutchinson in a report in *Compost Science* (52), discusses experiments conducted in Kenya on the production of methane gas from farmyard manure. Three different types of methane plants are described. The article also described the effects that the residue from the methane plants has had on 50 acres of coffee.

Methane plants have been designed to operate from grass, straw, coffee pulp and other organic material mixed with manure; and from manure mixed with water to make it into a liquid sludge. The gas produced by the plants is used as an energy source.

The running costs of the plants have been extremely small, and the labor used for running them is employed for other tasks as well. The plants are also able to produce fertilizer. Two of the plants discussed cost about 400 pounds sterling to install, and with coffee selling at about 350 pounds sterling per ton, the initial capital outlay is soon repaid.

Dick Shuttleworth, an Indiana farmer, has constructed a device that transforms ordinary cow manure into home fuel oil and gasoline (53). He has used this homemade fuel to run an automobile engine, power a gas stove and fuel lamps, operate a gas refrigerator, and fire a space heater.

This methane generator turns organic waste into both a nitrogen-rich fertilizer and methane gas. This design is so simple that almost any home craft handyman could build one of these generators. This generator should be as easy to install and operate as an air conditioning system.

Another home power unit has been designed and built by L. Auerback and B. Katz and is being demonstrated at the University of California at Berkeley (54). The system digests anaerobically human, pet and livestock wastes, as well as other animal and vegetable matter, to produce compost and a gas useful as a fuel source for cooking stoves, hot water heaters and internal combustion engines. In experiments with chicken manure, it was found that one ton of fresh manure mixed with small amounts of human waste produced more than 2,000 cubic feet of gas. A complete description of the system including construction and operating procedures is available for five dollars from Alternative Energy Systems, 242 Copse Road, Madison, CT 06443.

REFERENCES

(1) Miner, J.R., "Annual Review of Literature: Agricultural (Livestock) Wastes," *Jour. Water Poll. Control Fed.*, 40:1150-1158, 1968.
(2) Miner, J.R., "Annual Review of Literature: Agricultural (Livestock) Wastes," *Jour. Water Poll. Control Fed.*, 41:1169-1178, 1969.
(3) Jeffrey, E.A., Blackman, W.C. and Ricketta, R.L., *Aerobic and Anaerobic Digestion Characteristics of Livestock Wastes*, Eng. Ser. Bull. No. 57, Univ. of Missouri, Columbia, Mo., 1963.
(4) Taiganides, E.P. and Hazen, T.E., "Properties of Farm Animal Excreta," *Trans. Amer. Soc. Agr. Eng.*, 9:374-379, 1966.
(5) Miner, J.R., Lipper, R.I., Rina, L.R. and Funk, J.W., "Cattle Feedlot Runoff: Nature and Behavior," *Proc. 21st Ind. Waste Conf.*, Purdue Univ. Ext. Serv., 121:834-847, 1966.
(6) Grub, W., Albin, R.C., Wells, D.M. and Wheaton, R.Z., "Engineering Analyses of Cattle Feedlots to Reduce Water Pollution," *Trans. Amer. Soc. Agri. Engr.*, 12:490-492, 1969.
(7) Miner, J.R., *Odor from Livestock Production*, Dept. of Agr. Engr., Oregon State Univ., Corvallis, Oregon 1973.
(8) Taiganides, E.P. and White, R.K., "Mission Impossible: Control Odors in Poultry Production Units," *Proc. 2nd Nat'l Poultry Litter and Waste Management Seminar*, Texas A & M Univ., 1968, p. 10.
(9) Merkel, J.A., Hazen, T.E. and Miner, J.R., "Identification of Gases in a Confinement Swine Building Atmosphere," *Trans. Amer. Soc. Agr. Eng.*, 12:310-314, 1969.
(10) Lebeda, D.L. and Day, D.L., *Waste Caused Air Pollutants Are Measured in Swine Buildings*, Agric. Expt. Sta., Univ. of Illinois, Urbana, *Illinois Res. Bull.* No. 15, 1965.
(11) Day, D.L., "Liquid Hog Manure Can Be Deodorized by Treatment with Chlorine or Lime," Agric. Expt. Sta., Univ. of Ill., Urbana, *Illinois Res.*, 16:10-12, 1966.
(12) Hammond, C.W., Day, D.L. and Hansen, E.L., "Can Lime and Chlorine Suppress Odors in Liquid Hog Manure?" *Agr. Engr.*, 49:340-343, 1968.
(13) Schacht, C.J., "Development of Liquid-Manure-Handling Equipment," *Trans. Amer. Soc. Agr. Eng.*, 10:161-163, 1967.
(14) Decker, M.A., and Reed, C.H., "The Plow-Furrow Cover Method of Waste Disposal," *Proc. 10th Nat'l Pork Ind. Conf.*, Univ. of Nebraska, 1967.
(15) Dornbusch, J.M. and Andersen, J.R., "Lagooning of Livestock Wastes in South Dakota," *Proc. 19th Ind. Waste Conf.*, Purdue Univ. Ext. Ser., 177:317-325, 1964.
(16) Clark, C.E., "Hog Waste Disposal by Lagooning," *Jour. San. Eng. Div., Proc. Amer. Soc. Civil Engr.*, 91 (SA6):27-42, 1965.
(17) Hart, S.A. and Turner, M.E., "Lagoons for Livestock Manure," *Jour. Water Poll. Control Fed.*, 37:1578-1586, 1965.
(18) McCoy, E., "Lagooning of Liquid Manure (Bovine): Bacteriological Aspects," *Trans. Amer. Soc. Agr. Eng.*, 10:784-790, 1967.
(19) Taiganides, E.P., Baumann, E.R., Johnson, E.P. and Hazen, T.E., "Anaerobic Digestion of Swine Wastes," *Jour. of Agr. Engr. Res. (Brit.)*, 8:327-331, 1963.
(20) Dale, A.C., *Tentative Criteria for Design, Construction and Operation of the Batch Type Pasveer Oxidation Ditch System for the Treatment of Animal Wastes*, Dept. of Agr. Engr., Purdue University, Lafayette, Ind., 1967.

(21) Jones, D.D., Converse, J.C. and Day, D.L., "Aerobic Digestion of Cattle Waste," *Trans. Amer. Soc. Agr. Eng.*, 11:757-762, 1968.
(22) Hermanson, R.E., Hazen, T.E. and Johnson, H.P., "Biooxidation of Swine Waste by the Activated Sludge Process," *Trans. Amer. Soc. Agr. Eng.*, 12:342-353, 1969.
(23) Pratt, G.L., Hardness, R.E., Butler, R.J., Parsons, J.L. and Buchanan, M.L., "Treatment of Beef Cattle Waste Water for Possible Reuse," *Trans. Amer. Soc. Agr. Eng.*, 12:471-474, 1969.
(24) Bratzler, M.W., and Long, T.A., "Digestion of Hydrolyzed and Cooled Poultry Waste by Ruminants," *Jour. of Animal Sci.*, 27:1509-1516, 1968.
(25) Byrley, T., "Manure Dryers Opening Promising New Market," *Heating and Air Conditioning Contractor*, 59:56-58, 1968.
(26) Braga, A., "Research of the Survival of Salmonellae in Farmyard Effluent Used in a Fertilizing Irrigation Plant," *Water Poll. Abs. (Brit.)*, 40: No. 845, 1967.
(27) Novak, B. and Lobl, F, "The Complex Effect of Manures and Fertilizers on the Yield of Crops," *Biol. Abs.*, 48, 82605, 1969.
(28) Loehr, R.C., *Pollution Implications of Animal Wastes—a Forward Oriented Review*, Dept. of the Interior, FWPCA, Robert S. Kerr Water Res. Center, 1968.
(29) Muehling, A.J., *Swine Housing and Waste Management*, Agr. Eng. Publ. No. 873, Dept. Agric Eng., Univ. of Illinois, 1969.
(30) Dugan, G.L., Golueke, A.G., Oswald, W.J. and Rixford, C.E., *Photosynthetic Reclamation of Agricultural Solid and Liquid Wastes*, San. Eng. Res. Lab. SERL Report No. 70-1, Univ. of California, Berkeley, 1970.
(31) Verley, W.E. and Miner, J.R., "A Rotating Flighted Cylinder to Separate Manure Solids," ASAE Paper No. 73-470, *Trans. Am. Soc. Agr. Eng.*, 1973.
(32) Golueke, C.G., "Temperature Effects on Anaerobic Digestion of Raw Sewage Sludge," *Sewage and Industrial Wastes*, 30, (10), 1225-1232, 1958.
(33) Golueke, C.G., Oswald, W.J. and Gotaas, H.B., "Anaerobic Digestion of Algae," *Applied Microbiology*, 5, 47-55, 1957.
(34) Golueke, C.G. and Oswald, W.J., "Biological Conversion of Light Energy to the Chemical Energy of Methane," *Applied Microbiology*, 7, 219-227, 1959.
(35) Oswald, W.J. and Golueke, C.G., "Solar Power Via a Botanical Process," *Mechanical Engineering*, 86 (2), 40-43, 1964.
(36) Klein, S.A. and Golueke, C.G., "Anaerobic Digestion with Sewage Sludge," in *Comprehensive Studies of Solid Wastes Management*, First Annual Report, SERL Report 67-7, Sanitary Engineering Research Laboratory, University of California, Berkeley, May 1967.
(37) Klein, S.A. and Golueke, C.G., "Anaerobic Digestion" and "Economics of the Anaerobic Digestion Process," in *Comprehensive Studies of Solid Wastes Management*, Second Annual Report, SERL Report 69-1, SERL, Berkeley, January 1969.
(38) Klein, S.A. and Golueke, C.G., "Anaerobic Digestion" in *Comprehensive Studies of Solid Wastes Management*, Third Annual Report, SERL Report 70-2, SERL, Berkeley, June 1970.
(39) Dugan, G.L., Golueke, C.G. and Oswald, W.J., "Poultry Operation with an Integrated Sanitation Waste Materials Recycling System," *Journal of Water Pollution Control Federation*, 44 (3), 432-440, 1972.
(40) Chan, Deh Bin and Pearson, E.A., *Comprehensive Studies of Solid Wastes Management: Hydrolysis Rate of Cellulose in Anaerobic Fermentation*, SERL Report 70-3, SERL, Berkeley, October 1970.
(41) Klein, S.A. and Golueke, C.G., "Anaerobic Digestion of Solid Wastes," in *Comprehensive Studies of Solid Wastes Management*, Final Report, SERL Report 72-3, SERL, Berkeley, May 1972.
(42) Gotaas, H.B., *Composting - Sanitary Disposal and Reclamation of Organic Wastes*, WHO Monograph Series, No. 31, 1956.
(43) Golueke, C.G., "Composting Farm and Garden Wastes," *Vector Views*, 2, (12), 58-64, 1955.
(44) Singh, R.B., "The Bio-Gas Plant—Generating Methane from Organic Wastes," *Compost Science*, 13(1), 20-25, 1972.

(45) Patel, J.B. and Patel, R.B., "Biological Treatment of Poultry Manure Reduces Pollution," *Compost Science,* 12, (5), 18-21, 1971.
(46) *Livestock Waste Management and Pollution Abatement,* Proceedings of the International Symposium on Livestock Wastes, Ohio State University, April 1972, published by the American Society of Agricultural Engineers, Saint Joseph, Michigan, 1971 (360 pp).
(47) Loehr, R.C., "Treatment and Disposal of Animal Wastes," *Industrial Water Engineering,* 7, (11), 14-18, November 1970.
(48) Hart, S.A., "Digester Tests of Livestock Wastes," *Journ. W.P.C.F.,* 35, 748, 1963.
(49) Singh, R.B., "Building a Bio-Gas Plant," *Compost Science,* 13(2), 12-16, March-April 1972.
(50) Savery, C.W. and Cruzon, D.C., "Methane Recovery from Chicken Manure Digestion," *Journal of the Water Pollution Control Federation,* 44 (12), 2349-2354, December 1972.
(51) "Lebanon County Agents Produce Methane Gas from Animal Waste," *Pennsylvania Econotes,* 2(3), 8, December 1973.
(52) Jutchinson, T.H., "Methane Farming in Kenya," *Compost Science,* 30-31, November-December 1972.
(53) "Bio-Gas," *Compressed Air,* 79(2):14, February 1974.
(54) Auerbach, L., *A Homesite Power Unit: Methane Generator,* second edition, 1974.

INDUSTRIAL WASTE TREATMENTS

The sources of the material in this chapter are PB 231 149, PB 238 291 and PB 240 768. For a complete bibliography of these reports, see page 222.

TREATMENT OF PETROCHEMICAL INDUSTRY WASTEWATERS

The petrochemical industry is of considerable magnitude in the United States. It represents the backbone of the plastics and related materials segment of our economy. Listed below are the 1970 annual production rates for 10 of the more common petrochemical products as reported by the U.S. Tariff Commission:

Product	U.S. Annual Production, lb/yr
Alcohols (C_9 and less)	10,500,000,000
Formaldehyde	4,400,000,000
Vinyl and vinylidene resins	3,800,000,000
Elastomers	4,400,000,000
Ethylene	18,000,000,000
Ethylene glycol	3,000,000,000
Acetic acid	1,900,000,000
Acetic anhydride	1,500,000,000
Acetone	1,600,000,000
Adipic acid	1,100,000,000

Approximately 1% of the product is lost in the wastewater. The most serious obstacle to methane production from solid wastes is the cost of residue disposal. This is not a significant factor in methane production from the anaerobic treatment of petrochemical wastes because they are primarily soluble. Thus, no refractory solids fraction is present in the original wastewater. Microbial synthesis is relatively small in the anaerobic treatment process, minimizing excess sludge production which must be disposed of.

Since refractory solids in the wastewater are negligible, solids in the effluent are of microbial origin. Consequently solids detention times in the order of hundreds of days are easily achieved. Likewise, efficient solids removal can be achieved by upflow granular beds. Methane conversion rates in excess of 1,000 lb of chemical oxygen demand equivalent per 1,000 ft^3 per day have been achieved. This, of course, results in considerably reduced size of the installation.

Quite often, the temperature of the petrochemical wastewaters is 80° to 100°F requiring no heat addition to the digester. Mixing energy requirements for digestion of petrochemical wastes are much reduced compared to solid wastes because the volume of the reactor and the solids concentration within the reactor are lower. Energy for shredding is not needed with soluble petrochemical wastes. Consequently, the net input of energy per 10^6 Btu of methane produced is considerably less for anaerobic digestion of petrochemical wastes versus solid wastes.

Heavy metals in the wastewaters tend to stay in solution in aerobic treatment processes resulting in toxicity to the biomass. Concentrations of 1 to 2 mg/l of some heavy metals can knock out an aerobic process resulting in loss of treatment capability. In the anaerobic process, sulfides are present or can be added which precipitate the heavy metals, thus removing them from the wastewater stream and eliminating their toxicity to the biomass. In many situations, this inherent ability to remove heavy metals is the deciding factor which makes anaerobic treatment the design choice.

A comparison of anaerobic and aerobic treatment per ton of biological oxygen demand per day is as follows (costs given in 1973 dollars):

	Aerobic	Anaerobic
Aeration costs ($)	20	0
Sludge disposal costs ($)	15	0
Area required (ft^2)	2,000	1,000
Heavy metals	Toxic	No effect
Methane	0	-$6 (credit)
Volume required (ft^3)	20,000	10,000

Thus, not considering the savings in volume and area requirements and the elimination of heavy metal toxicity and its removal from the wastewater for a typical plant which will produce in the order of 30 tons of biological oxygen demand per day for 200 days per year, approximately $250,000 could be saved per year using anaerobic instead of aerobic treatment. Thus, it may prove to be very advantageous, costwise, to segregate certain wastewater streams in a plant, so that anaerobic treatment can be utilized.

RUM DISTILLERY SLOPS TREATMENT

The production of rum is accomplished by age-old processes that, in their modern application, are similar in function throughout the industry. The basic sequence of steps in rum production consists of: the mixing of molasses, water, nutrients, and antifoamant; acidification of the above mixture; propagation of the biomass used in the fermentation; the fermentation itself; and the distillation of the ferment.

The distilled spirits are transferred sequentially to holding tanks, to oak barrel filling, to the aging warehouse, and (after a legally-defined period of aging) to barrel emptying operations and thence to the bottling line, in the typical operation. The above rum production and bottling operations are often conducted on a seasonal basis, with shutdown periods of one to two months each year for equipment maintenance and replacement and other reasons.

The basic types of process waste streams generated in the production of rum include:

(1) The slops stream, i.e., the underflow produced in the distillation of the fermented molasses mixture.
(2) Barrel washings.
(3) Cooling tower and boiler plant blowdown.
(4) Regenerant water from water treatment facilities, analytical laboratory wastewaters, and fermenter washdown.

Presented in Table 6.1 is a profile of (1) the total facility wastewater generation per proof gallon produced from all of the above process streams, and (2) an analysis of the percent contribution to the total wastewater generation by each type of process stream. In review of these data:

(1) The slops discharge constitutes 66% of the waste flow, over 98% of the BOD and COD emissions, over 90% of the solids emissions, and essentially all of the nitrogen and phosphorus emissions in rum production operations.
(2) The second most important wastewater source on a flow volume basis is the boiler and cooling water blowdown/fermenter washdown stream, the flow component of which is derived primarily from the blowdown streams, and the organic component of which is derived from the fermenter washdowns.

TABLE 6.1: WASTEWATER GENERATION IN RUM PRODUCTION

Waste Parameter or Constituent	Total Facility Waste Generation per Proof Gallon	Contribution by Type of Waste Stream			
		Slops Stream (%)	Barrel Washings (%)	Boiler/Cooling Water & Fermenter Washdown (%)	Water Treatment & Analytical Lab. Wastewaters (%)
Volume	55.6 l (14.7 gal)	66	5	26	3
COD	3.0 kg (6.6 lb)	98	1	1	–
BOD	1.0 kg (2.3 lb)	99	–	1	–
Total solids	4.2 kg (9.2 lb)	91	–	9	–
Total dissolved solids	3.9 kg (8.6 lb)	91	–	9	–
Total suspended solids	0.25 kg (0.56 lb)	97	–	3	–
Total Kjeldahl nitrogen	0.06 kg (0.14 lb)	100	–	–	–
Total phosphate	0.003 kg (0.007 lb)	100	–	–	–

Source: PB 238 291

It is apparent from the preceding that the slops stream constitutes the major wastewater management problem in this industry.

The slops stream typically has the following characteristics:

(1) COD—70 to 100 g/l
(2) BOD—20 to 60 g/l
(3) Total suspended solids—3 to 10 g/l
(4) Total dissolved solids—75 to 85 g/l
(5) Total nitrogen—0.8 to 1.5 g/l
(6) Total phosphorus—60 to 100 mg/l
(7) Sulfate—3 to 5 g/l
(8) pH—4.0 to 4.7
(9) Color—100,000 units
(10) Temperature—80° to 90°C

The principal factors associated with the magnitude and variation of these characteristics are: the variable sugar and ash contents of the molasses, which itself is a by-product of sugar production; and the amount of acidification (H_2SO_4) of the molasses-water mixture to obtain an optimal pH level for the fermentation. The slop is rich in nutrient materials having value as cattle feed extenders and soil supplements. From a biological treatment perspective, the slops stream typically contains a deficiency of nitrogen and phosphorus, and most of the volatile suspended solids component of the slops stream is derived from the yeast crop produced in the fermentation.

In recognition of the significance of the slops stream as a wastewater management problem for the rum industry, a study was initiated to examine the efficacy of an anaerobic biological treatment system as the initial processing step in the treatment of slops. There are several basic reasons for selecting anaerobic biological treatment for examination, the foremost of which are:

(1) The organic content of the slops (70 to 100 g/l COD) is well in excess of the point where the oxygen demand in an aerobic system exceeds economically attainable rates of oxygen transfer.

(2) The anaerobic fermentation process is characterized by low kinetic reaction constants and biomass yields, the intramolecular breakdown of complex organic compounds, and the production of methane gas as a by-product.

(3) Sludges produced in a properly functioning anaerobic fermenter are generally more easily dewatered than are aerobic sludges, and a lesser mass of waste solids is produced per unit volume of wastewater treated in the anaerobic process than in the aerobic process.

A study was initiated in May 1972 with the development of a laboratory facility and pilot plant site at the Palo Seco, Puerto Rico distillery of Bacardi Corporation. During the latter half of 1972:

(1) Two bench-scale anaerobic contact process units were designed and made operational.

(2) Pilot plant facilities and instrumentation were leased from the University of Puerto Rico, and refurbished to incorpor-

Industrial Waste Treatments

ate a separate clarifier/thickener for use (in an anaerobic contact process configuration) in the study.

(3) Work plans and preliminary reports for the investigations were developed.

(4) Seed culturing techniques were established and seed cultures were developed for use in the startup of each of the three units.

Biological treatability studies were initiated with the startup of the two bench-scale systems in February 1973, and the pilot plant in April 1973. The bench-scale units were operated over a nine-month period under a wide range of hydraulic and organic loading rates, to develop the information necessary to establish the process kinetics (Monod kinetics) describing the anaerobic biological treatment of slops in the anaerobic contact process. The pilot plant operations were designed to develop the information necessary to verify the process kinetic parameters at the pilot-scale, and to investigate the settling characteristics of mixed liquor solids suspensions.

During the last month of pilot plant operations (December 1973), the pilot plant was modified to permit evaluation of the settling characteristics of the mixed liquor suspended solids under a wide range of surface loading rates and sludge recycle rates, both with and without polymer preconditioning of the mixed liquor solids suspension; the results of this terminal activity in the study were used to develop the design criteria for the gravity solids separation components of the full-scale application.

Status of Technology of Rum Distillery Slops Digestion

Much of the experience on the treatment of spent molasses wastes by the anaerobic digestion process has been summarized in an early (1959) review by Pettet et al (1) and in a later review by Hiatt et al (2). Two factors reported by Pettet et al (1) as interfering in the anaerobic digestion of rum distillery slops are the high sulfate concentrations in the slops stream, and the production of volatile acids in high concentrations during the anaerobic digestion. To circumvent the former concern (sulfide inhibition as a result of sulfate reduction to sulfide) and the general problem of high salinity concentrations in slops, much of the research to date has been conducted using diluted spent molasses (slops) waste feed streams.

Stander (3) was able to obtain a 66% removal of organic carbon from a molasses slop diluted with an equal volume of water at a hydraulic residence time of 3.75 days, an organic loading of 8.8 kg volatile matter/day/m^3 (0.55 lb volatile matter/day/ft^3), and at 33°C. In this study a repeated reinoculation technique, equivalent to a form of solids recycle, was necessary to sustain the fermentation. Hiatt et al (2), reported that a stable fermentation was possible with a diluted slops stream (65% of full strength) at a hydraulic residence time of 8.4 days, an organic loading of 7.7 kg total (unfiltered sample) COD/day/m^3 (0.48 lb total COD/day/ft^3), and at a temperature of 35°C.

A 71% removal of total COD was obtained. These researchers also reported that a sustained fermentation of full strength slops was obtained at a hydraulic residence time of 16 days, and at an organic loading rate of 5.9 kg total COD per day per cubic meter (0.37 lb total COD/day/ft^3), with a resultant COD removal of 70%. After initial stabilization of the digesters, the gas produced contained about 65% methane, and the methane in the gas production represented about 69% of the COD removal. These researchers suggested that gas scrubbing to remove hydrogen sulfide and subsequent recirculation through the fermenter would be an effective control action to relieve inhibition due to sulfide ion.

The observations of Hiatt et al (2) that high volumetric loadings are possible if the waste is diluted confirmed earlier work on the anaerobic digestion of rum distillery slops by Radhakrishnan et al (4). In laboratory studies conducted at 37°C, these investigators found that the maximum possible loading with undiluted waste was 3 kg BOD/day/m^3 (0.18 lb BOD/ft^3/day) at a hydraulic residence time of 11 days, whereas a maximum loading of 4.5 kg BOD/day/m^3 (0.28 lb BOD/ft^3/day) was attained at a five-day hydraulic residence time with a diluted waste (60% of full strength).

BOD removals in these studies averaged from 80 to 85%. The gas contained from 54 to 60% methane, and more than 1% hydrogen sulfide. In evaluating results, it was reported that volatile acids levels in excess of 2,000 mg/l (as acetic acid) did not impair the fermentation, providing that a sufficiency of alkalinity was present to maintain the pH of the system in excess of 7.0. Additionally, sulfide concentrations as high as 240 mg/l were not found to be inhibitory to the digestion of undiluted wastes.

Process Flow Sheet and Design Criteria

The process flow sheet and design criteria for full-scale rum distillery slops treatment by the anaerobic contact process were developed in consideration of the evaluations and the base of operating experience accrued over a two-year period of investigation with bench-scale and pilot plant anaerobic contact units. The process flow sheet is presented in Figure 6.1, and the recommended design criteria are summarized in Table 6.2 and discussed below.

Unit Processes/Operations: The function of the slops storage tank is to provide equalizing of hour-to-hour flow variation, and to permit the cooling of the slops stream emanating from the distillation columns to a temperature of 32° to 38°C. The excess sensible heat in this tank can be used to maintain the digester temperatures at 35 ± 2°C, as an option to using a portion of the methane gas production for this purpose. A minimum storage capacity of 1 m^3/m^3/day of slops flow (1,000 gal/1,000 gpd slops flow) is suggested, exclusive of situation-specific storage requirements for restartup should the latter be required.

The digester capacity required at the selected hydraulic residence time of 15 days is 15 m^3/m^3/day of flow (2,000 ft^3/1,000 gpd), and a minimum of two digesters should be provided for operational flexibility. The digesters should be heated to 35 ± 2°C. A gas mixing capacity of 0.15 m^3/min/m^3 of digester volume (20 scfm/1,000 gal) is suggested as a criterion for full-scale design based on the satisfactory pilot plant experience at this level.

The clarifiers should be sized to provide for a solids loading rate of 100 kg/day/m^2 (20.5 lb/day/ft^2) which, at an MLTSS concentration of 10,320 mg/l, is equivalent to a liquid loading rate (overflow basis) of 9.69 m^3/day/m^2 (238 gpd/ft^2). A recycle ratio of at least 2:1 (ratio of slops feed flow:sludge recycle flow) is

Industrial Waste Treatments

FIGURE 6.1: PROCESS FLOW SHEET—RUM DISTILLERY SLOPS TREATMENT BY ANAEROBIC CONTACT PROCESS

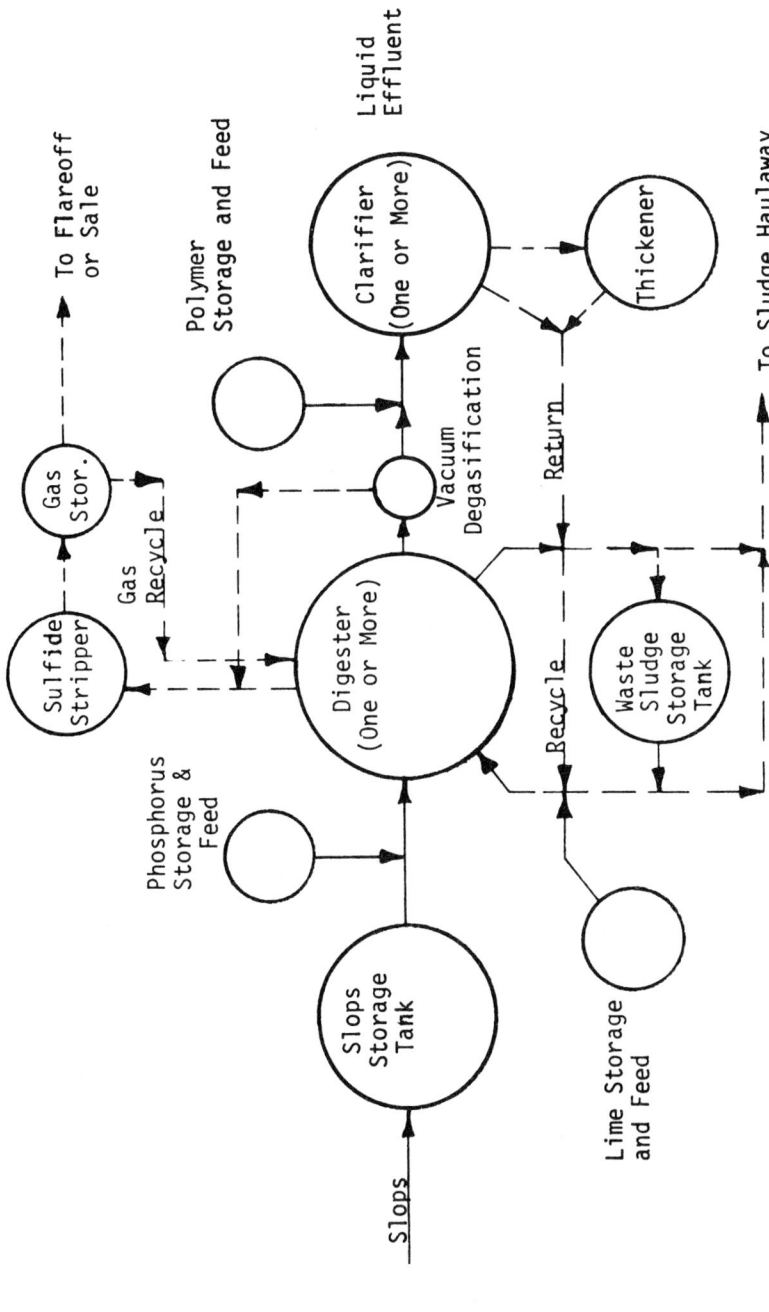

Source: PB 238 291

recommended based on the experience of Steffen and Bedker (5) in a slaughterhouse waste/anaerobic contact process application. For a recycle ratio of 2:1, the TSS concentration in the recycle stream is equal to 13,550 mg/l, and the design liquid loading rate (total basis) is equal to 29 m^3/day/m^2 (713 gpd/ft^2). A design weir loading rate not to exceed 12.4 m^3/day/m (1,000 gpd/ft) is recommended on the basis of pilot plant experience.

TABLE 6.2: DESIGN CRITERIA FOR ANAEROBIC CONTACT PROCESS

Process Element	.. Recommended Design Criteria (units as noted) ..	
Slops storage tank	1 m^3/m^3/day slops flow	1,000 gal/1,000 gpd slops flow
Digesters		
Capacity	15 m^3/m^3/day slops flow	2,000 ft^3/1,000 gpd slops flow
Gas mixing	0.15 m^3/min/m^3 digester capacity	20 scfm/1,000 gal digester volume
Heating	35±2°C	95±3°F
Clarifiers		
Liquid loading rate (overflow basis)	9.69 m^3/day/m^2	238 gpd/ft^2
Weir loading rate	12.4 m^3/day/m	1,000 gpd/ft
Recycle ratio	2:1	2:1
Thickeners		
Dry solids loading rate	122 kg/day/m^2	25 lb/day/ft^2
Waste sludge storage tank	0.52 m^3/m^3/day slops flow	70 ft^3/1,000 gpd slops flow
Gas handling system	36 m^3 gas/m^3/day slops flow	4,750 ft^3 gas/1,000 gpd slops flow
Sulfide stripper	0.43 kg S/m^3/day slops flow	3.6 lb S/1,000 gpd slops flow
Chemical feed systems		
Lime	4 kg CaO/m^3/day slops flow	33.3 lb CaO/1,000 gpd slops flow
Phosphorus	0.3 kg P/m^3/day slops flow	2.5 lb P/1,000 gpd slops flow
Polymer	15 g/m^3/day slops flow	0.125 lb/1,000 gpd slops flow

Source: PB 238 291

In the specification of criteria for the thickener, it was assumed that: (1) the settled solids can be thickened at a loading rate of 122 kg/day/m^3 (25 lb dry solids/day/ft^2; and (2) the thickened sludge will have a dry solids concentration of 10%. These values were selected on the basis of design parameters typically associated with the thickening of primary sludges in municipal wastewater treatment plants.

The functions intended for the waste sludge storage are to: (1) provide waste sludge storage for a reserve sludge quantity equal to one-third of the biomass inventory in the digesters, to be available for transfer to the digesters as needed; and (2) to provide interim sludge storage prior to haulaway. The MLVSS biomass inventory required (103.7 kg MLVSS/m³/day of slops heated), at a volatile fraction of 0.67, is equivalent to an MLTSS inventory of 155 kg MLTSS/m³/day.

For the storage of one-third of this inventory in the waste sludge storage tank, at an average concentration of 10% solids, the storage capacity required is equal to 0.5 m³/m³/day of slops treated (70 ft³/1,000 gpd). At a waste MLTSS production rate 3.87 kg/day/m³/day, the waste sludge storage capacity provided is sufficient to store 13 days of waste sludge production.

The gas handling system should be sized to handle 36 m³ of gas production per cubic meter per day of slops treated. Assuming that the gas stream contains 0.8% hydrogen sulfide by volume, then the sulfide stripper should be sized to handle 0.43 kg elemental sulfur/m³/day of slops flow (3.6 lb/1,000 gpd of slops flow).

Chemical feed systems are required for the addition of lime, phosphorus and polymer. Based on observations, it is suggested that these systems be sized as follows:

(1) Lime—feed at 4 kg as CaO/m³/day of slops flow (33.3 lb/1,000 gpd)
(2) Phosphorus—feed at 0.3 kg as P/m³/day of slops flow (2.5 lb as P/1,000 gpd)
(3) Polymer—feed at 15 grams polymer/m³/day of slops flow (0.125 pounds polymer/1,000 gpd)

Layout: The layout as exemplified by the process flow sheet of Figure 6.1 incorporates the following material/liquid transfer capabilities:

(1) From either digester to another or the same digester,
(2) From the waste sludge storage tank to the digesters or to sludge haulaway,
(3) From the digesters to the clarifier,
(4) From the clarifier to the thickener, the waste sludge storage tank, and/or directly to the digesters.

The above transfer capabilities are required to support the types of material/liquid transfers required in the startup, routine operation, and restartup phases as discussed earlier in this section.

Economic Analysis

The economic analysis was developed using the following approach:

(1) The design criteria of Table 6.2 and the process flow sheet of Figure 6.1 were used to develop estimates of capital and operating/maintenance costs at two design flow rates, 190 and 1,140 m³/day (50,000 and 300,000 gpd);
(2) The cost estimates were developed using available unit process cost curves, updated to January 1974;
(3) The heating requirements for the digesters at each design flow were estimated in consideration of ambient conditions in Puerto Rico, wherein most of the North American rum distilling capacity is located;

(4) The value of the energy available as methane by-product was assumed to be equal to the cost of purchasing an equivalent amount of energy in the form of fuel oil; and

(5) The cost estimates at each design flow rate were converted to unit treatment costs ($/vol treated) from which a cost capacity relationship was then constructed.

Capital Costs: The estimated capital costs for anaerobic contact treatment systems at design capacities of 190 m^3/day (50,000 gpd) and 1,140 m^3/day (300,000 gpd) are presented in Table 6.3. These cost estimates were developed from cost data presented by Eilers and Smith (6) and Patterson et al (7), updated to January 1974 cost levels, and by vendor contact. Overhead (engineering and administration) and contingency costs were estimated at 38% of the construction cost.

TABLE 6.3: CAPITAL COSTS (JANUARY 1974 DOLLARS)

Item	Capacity at 190 m^3/day (50,000 gpd)	Capacity at 1,140 m^3/day (300,000 gpd)
Slops storage tank*	32,000	112,300
Mechanical mixers—slops storage tank	12,000	42,000
Control and solids handling building**	175,000	634,000
Lime storage, slaking, and feeding system	21,000	78,000
Phosphorus storage, dissolution and feeding system	3,000	10,000
Polymer storage and feeding system	1,500	5,000
Sulfide stripper	70,000	200,000
Gas storage/handling	90,000	270,000
Anaerobic digesters***	400,000	1,400,000
Vac. degasification	50,000	170,000
Gravity separator	65,000	125,000
Gravity thickener	38,000	80,000
Waste sludge storage	8,600	28,000
Pumps	12,500	46,000
General work site work, yard piping, general elec. and HVAC	240,000	855,000
Construction cost	1,218,600	4,055,300
Overhead; contingency	463,200	1,541,000
Total construction cost	1,681,800	5,596,300

*Concrete in place at $275/yd^3
**Includes areas for control room, lime storage, slaking, feeding, sulfide stripper and gas handling equipment, phosphorus storage, dissolution and feeding equipment, polymer storage, dissolution and feed equipment, pumps, etc.
***Concrete at $275/yd^3; covers; heat exchanger

Source: PB 238 291

The estimated total construction costs were $1,682,000 for a 190 m³/day facility, and $5,596,000 for a 1,140 m³/day facility. The corresponding unit total construction costs are: $8,850/m³/day of capacity ($33,660/1,000 gpd) at the design flow rate of 190 m³/day; and $4,910/m³/day ($18,650/1,000 gpd) at the design flow rate of 1,140 m³/day.

Operating and Maintenance Costs: The estimated annual O/M (operating and maintenance) costs for anaerobic contact systems at design capacities of 190 m³ per day and 1,140 m³/day are presented in Table 6.4. Labor costs were estimated at a rate of $5.40/man hour, including overhead. The chemical costs were estimated as follows: lime at $17.50/ton; phosphorus at $50/ton; and polymer at $1/lb. Sludge haulaway costs were estimated at a unit cost of $0.30/ton dry solids/mile of haul, assuming a 25 mile haul. Electricity costs were estimated by assuming a pumping requirement of 0.004 hp/1,000 gpd of slops flow (6), and an annual cost of $330/yr/hp, the latter based on pump efficiencies of 60% and electricity costs of $0.03/kwh.

TABLE 6.4: ANNUAL OPERATING AND MAINTENANCE COSTS (JANUARY 1974 DOLLARS)

Item	Capacity @ 190 cu m/day (50,000 gpd)	Capacity @ 1,140 cu m/day (300,000 gpd)
Labor	16,200	54,000
Materials and supplies (except chemicals)	8,000	20,000
Chemicals	6,000	35,400
Electricity	6,800	27,000
Other utilities	500	2,000
Sludge haulaway	2,000	12,000
Total O/M Cost[a]	39,500	150,400

Note: [a] excludes heating cost.

Source: PB 328 291

Fuel costs for digester heating are not included in Table 6.4 on the assumption that a portion of the methane production would be allocated for this purpose. An evaluation of the allocation required, and of the net methane production after the allocation is made, is presented below.

Exclusive of fuel cost for heating, the estimated annual O/M costs are $39,500 at the design capacity of 190 m³/day (50,000 gpd), and $150,400 at the design capacity of 1,140 m³/day (300,000 gpd).

Economic Value of Methane Production: The factors considered in determining the economic value of the methane production were:

(1) The gross methane production;
(2) The allocation required, from the gross methane production, to heat digesters;
(3) The net methane production, i.e., the excess available after deduction of the digester heating requirement from the gross methane production.

The first step in the evaluation was the estimation of digester heating requirements at each scale. The requirements were estimated using the procedure of Babbitt and Baumann (8) and the following assumptions (assuming location of the facilities in Puerto Rico):

(1) Average ambient temperature: 21°C (70°F)
(2) Digester feed stream and sludge recycle stream temperatures: 27°C (80°F)

On this basis, the estimated Btu requirements for digester heating were: 2.45×10^9 Btu/yr at the design capacity of 190 m^3/day; and 13.8×10^9 Btu/yr at the design capacity of 1,140 m^3/day. Given that a barrel of fuel oil has a Btu equivalent of approximately 6,000,000 Btu, the preceding Btu requirements for digester heating are equivalent to fuel oil consumption rates of: 408 bbl/yr at a design capacity of 190 m^3/day; and 2,300 bbl/yr at a design capacity of 1,140 m^3/day.

The fuel oil equivalent of the gross annual methane production was estimated using the following:

(1) The gross methane production factor of 21.3 m^3 CH$_4$/m^3 slops treated (2,850 ft^3/1,000 gal treated);
(2) An energy equivalent of 1,000 Btu/ft^3 of methane, or 5.90 bbl of fuel oil/1,000 m^3 of methane (0.167 bbl of fuel oil/1,000 cubic feet of methane);
(3) An assumed annual production schedule of 300 days.

For these conditions, the fuel oil equivalent of the gross annual methane production is 37.7 bbl/yr/m^3/day of capacity (143 bbl/yr/1,000 gpd); i.e., is equal to 7,160 bbl/yr at a design capacity of 190 m^3/day (50,000 gpd), and 42,900 barrels per year at design capacity of 1,140 m^3/day (300,000 gpd).

The allocation of gross methane production required for digester heating purposes can be evaluated by comparing the ratio of fuel oil equivalent consumption for digester heating with the above fuel oil equivalent production rates, at each design scale. This ratio is equal to 408/7,160 (5.7%) at 190 m^3/day capacity, and 2,300/42,900 (5.4%) at 1,140 m^3/day capacity. Thus, at either scale, less than 6% of the gross methane production must be allocated for digester heating in the Puerto Rico environment, and the net methane production is equal to at least 94% of the gross methane production. At 94%, the fuel oil equivalent of the net methane is equal to 35.4 bbl/yr/1,000 gpd. On a dollar basis, this fuel equivalent is worth $0.119/m^3 of slops treated (0.45/1,000 gal treated) for each $/bbl of fuel oil cost on the open market.

Treatment Costs: Annual treatment costs ($/m³ or $/1,000 gal of slops treated) were estimated on an unadjusted basis (assuming a fuel oil value of $0/bbl) and on an adjusted basis assuming fuel oil values up to $28/bbl. A summary of the unadjusted annual costs at each design flow rate is presented in Table 6.5. The amortization costs were developed assuming an interest rate of 8%, a 15 year life for 40% of the capital investment, and a 25 year life for 60% of the capital investment.

TABLE 6.5: UNADJUSTED ANNUAL COSTS FOR RUM DISTILLERY SLOPS TREATMENT BY ANAEROBIC CONTACT PROCESS

Item	Capacity @ 190 cu m/day (50,000 gpd)	Capacity @ 1,140 cu m/day (300,000 gpd)
Capital costs[1]	$1,683,200	$5,596,300
Annual costs ($/yr)		
. Amortization[2]	173,269	576,083
. O/M costs[3]	39,500	150,400
. Total annual cost[4]	212,769	726,483
Unit treatment costs[5]		
. Per cu m treated	3.74	2.13
. Per 1,000 gallons treated	14.18	8.07

[1]From Table 6.3

[2]Amortization at 8% interest rate; 15 year life for 40% of capital investment and 25 year life for 60% of capital investment

[3]From Table 6.4

[4]Exclusive of methane credit

[5]Based on 300 day production schedule

Source: PB 238 291

The unadjusted total annual cost of the full-scale applications is $212,800 at the design capacity of 190 m³/day (50,000 gpd), and $726,500 at 1,140 m³/day (300,000 gpd). For a 300 day production schedule, the unadjusted total annual costs are equivalent to unit treatment costs of: $3.74/m³ treated ($14.18 per 1,000 gal treated) at the design capacity of 190 m³/day; and $2.13/m³ treated ($8.07/1,000 gal treated) at the design capacity of 1,140 m³/day.

The unadjusted unit treatment cost data of Table 6.5 were used to construct the cost vs capacity curve of Figure 6.2 at the fuel oil value of $0/bbl. The adjusted unit treatment cost curves of Figure 6.2, at fuel oil prices of $4 to

FIGURE 6.2: UNIT TREATMENT COSTS FOR RUM DISTILLERY SLOPS TREATMENT BY ANAEROBIC CONTACT PROCESS

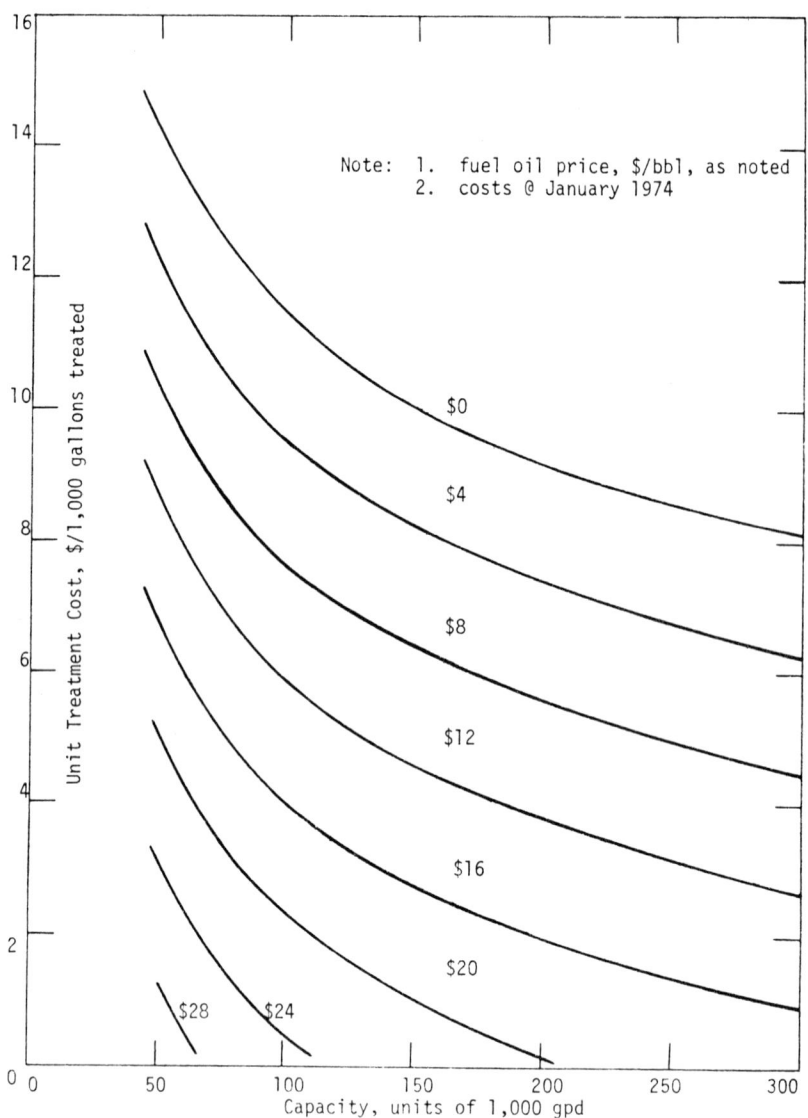

Source: PB 238 291

$28/bbl in $4/bbl increments, were developed by deducting $1.80/1,000 gal treated for each $4 increment, from the unadjusted costs.

The significance of design capacity and fuel oil price on the unit cost of rum distillery slops treatment by the anaerobic contact process is clearly evident from the cost curves of Figure 6.2. The unit treatment cost (in units of $/1,000 gallons treated) is $0.80 less in a 100,000 gpd facility than in a 50,000 gpd facility and an additional $3.30 less in a 300,000 gpd facility than in a 100,000 gpd facility, for any given fuel oil price.

If it is assumed (conservatively) that fuel oil prices will soon average $12/bbl, then the incorporation of methane by-product recovery in a plant-scale installation can reduce unit treatment costs from 35% (in a 50,000 gpd facility) to 65% (in a 300,000 gpd facility), as compared with unit treatment costs for installations without recovery.

UTILIZATION OF BIOGAS IN SUGAR INDUSTRY

The biogas-biomanure process, developed at the National Sugar Institute in collaboration with Hungarian experts, is described by S.C. Gupta, J.P. Shulka and K.A. Prabhu in the December 1967 issue of *Indian Sugar* (9). In this process, bagasse, cane trash, press mud, boiler ash, and cane yard sweepings can be actively fermented under controlled conditions resulting rapidly in a combustible gaseous mixture of 65% methane and 35% carbon dioxide called biogas. In addition, a manure is formed which is rich in nutrients and humus, and which has received favorable response compared to fertilizer or farm yard manure.

The advantages of this process are: the two major wastes, cane trash and bagasse, along with other sugar factory wastes, are properly utilized; their fuel value does not get lost and alternate fuel is avoided by burning of biogas, with an efficiency of over 85%; and the return of biomanure to cane areas results in increased yield and better economy. Since the presence of carbon dioxide lowers the calorific value of biogas, attempts have been made to utilize the carbon dioxide in the preparation of precipitated chalk for paper factories and for cane juice clarification. These processes are described and the tabulated results show that the carbon dioxide in the biogas can be successfully used in these processes.

TREATMENT OF WINERY WASTES

According to a report given at the 22nd Industrial Waste Conference at Lafayette, Indiana, May 2-4, 1967, a full-scale plant was constructed by the National Institute for Water Research of Pretoria, South Africa, to investigate the treatment of wine distillery wastes (10). The plant used was a modified Dorr-Oliver Clarigester of 40-ft diameter, with a digester compartment of 23,000 ft^3. A diagram of the full-scale plant was presented. Digester performance, load rates, temperature, sludge metabolism, digester solids concentration, sludge stabilization, volatile fatty acids, and nitrogen content of digester sludge are discussed and data were tabulated.

Process control evaluations included: effect of temperature, sludge density, availability of nutrients, and continuous digestion and sludge stability. It was estimated that anaerobic treatment of spent wine would amount to 56 cents (1 South African cent = 1.4 U.S.A. cents) per 1,000 gal. The modified Clarigester unit is easily controlled and operated by normal sewage works personnel. Proper acclimatization of sludge eliminated the need for reinoculation practice and rendered the sludge resistant to shock loads and sudden adverse temperature changes. The purification afforded was consistently above 96%. The gas liberated from the digestion process amounted to 14 v/v of spent wine and constitutes a potential heat source.

REFERENCES

(1) Pettet, A.E., Tomlinson, T.G., and Hemens, J., "The Treatment of Strong Organic Wastes by Anaerobic Digestion," *Journal Institution of Public Health Engineers,* 170, July 1959.

(2) Hiatt, W.C., Carr, A.D., and Andrews, J.F., "Anaerobic Digestion of Rum Distillery Wastes," presented at 28th Industrial Waste Conference, Purdue University, May 1973.

(3) Stander, G.J. and Snyders, R., "Effluents from Fermentation Industries, V," *Journal Institute of Sewage Purification,* Part IV, 447, 1950.

(4) Radhakrishnan, I., De, S.B., and Nath, B., "Evaluation of the Loading Parameters for Anaerobic Digestion of Cane Molasses Distillery Waste," *Journal Water Pollution Control Federation,* 41:R431, 1969.

(5) Steffen, A.J., and Bedker, M., "Operation of Full-Scale Anaerobic Contact Treatment Plant for Meat Packing Wastes," *Proceedings of 16th Industrial Waste Conference,* Purdue University, 1961. Ext. Ser 109, 423.

(6) Eilers, R.G., and Smith, R., *Wastewater Treatment Plant Cost Estimating Program,* U.S. Environmental Protection Agency, Cincinnati, Ohio, April 1971.

(7) Patterson, W.L., and Barker, R.F., *Estimating Costs and Manpower Requirements for Conventional Wastewater Treatment Facilities,* Environmental Protection Agency Project #17090 DAN, October 1971.

(8) Babbitt, H.E., and Bauman, E.R., *Sewage and Sewage Treatment,* New York, John Wiley and Sons, Eighth Edition, 1958, 598-9.

(9) Gupta, S.C., Shulka, J.P., and Prabhu, K.A., "Utilization of Biogas in Sugar Industry," *Indian Sugar,* vol 17, no. 9, 675-676, December 1967.

(10) "Treatment of Wine Distillery Wastes by Anaerobic Digestion," *Proceedings 22nd Industrial Waste Conference,* Lafayette, Indiana, May 2-4, 1967, Purdue University Engineering Extension Series, No. 129, 896-907.

METHANE FROM ENERGY CROPS

The sources of the material in this chapter are the following reports:
PB 231 149, PB 241 055 and PB 238 103. For a complete bibliography, see
p 222.

The conversion of waste materials to fuel gas via anaerobic digestion represents a potential solution not only to the energy problem, but simultaneously to the waste disposal problem. However, for the long range, the cultivation of energy crops will be required for the production of the organic matter needed to fuel the digesters. The available options include the cultivation of terrestrial, freshwater or marine plants. These are termed "energy crops" because they are grown solely to convert the energy of the sun, via photosynthesis, to plant material which can be anaerobically digested to produce methane. This chapter describes some of the work being done on the bioconversion of these plant materials.

STUDIES ON ALGAE AT UNIVERSITY OF CALIFORNIA (BERKELEY)

At the Bioconversion Energy Research Conference of June 1973, C.G. Golueke of the Sanitary Engineering Research Laboratory (SERL) at the University of California (Berkeley) summarized studies done on algae at SERL. According to Golueke, the work on the conversion of solar energy to the chemical energy of methane by way of algae culture and anaerobic digestion of the algae so produced was a by-product of extensive studies carried on at SERL since 1950 on the use of photosynthesis for the reclamation of water and nutrients from wastewaters.

The research on waste treatment until the year 1965 is fairly well summarized in the paper "Harvesting and Processing of Sewage Grown Planktonic Algae" (1). The work on the conversion of solar energy to methane was begun with a study on the anaerobic digestion of algae (2). The culmination was the development of a partially closed system for converting light energy to the chemical energy of methane through the agency of unicellular algae and aerobic and anaerobic bacteria operating in an integrated unit consisting of an algal growth unit (oxygenation chamber), "activated sludge" unit, and digester (3).

Digestion of Algae

The investigation of algae involved: (a) the making of a comparison of the value of raw sewage sludge with that of plankton green algae as a source of nutrient in anaerobic digestion; and (b) the determination of the effect of temperature, of aluminum in the algal feed, of detention period, and of loading on digester performance. The source of the aluminum in the feed was the aluminum sulfate used in harvesting the algae.

The experimental conditions were as follows: (1) the digesters were operated at 30° and 50°C, (2) the algal feed came from outdoor high-rate sewage treatment ponds containing an algal population composed mostly of Scenedesmus sp. (about 80%) and Chlorella sp. (about 20%) (the algal concentration of the ponds varied from 200 to 400 mg/l, dry weight), (3) the algae were harvested either by centrifugation or by precipitation with aluminum sulfate.

The volume of gas produced from the digestion of the raw sludge averaged 16.3 cubic feet per pound of volatile matter destroyed; while that from the digestion of algae averaged 15.8 cubic feet. Gas production per pound of volatile matter introduced averaged 9.5 cubic feet when raw sewage sludge was digested; and depending upon temperature, ranged from 6.1 to 8.0 cubic feet when algae (centrifuged) were digested.

Gas production per pound of volatile matter introduced was consistently lower at 35°C than at 50°C. The difference is a reflection of the difference in percentage destruction of volatile matter introduced, which amounted to 54% at 50°C and from 36 to 44% at 35°C. (Destruction of sewage sludge volatile matter averaged 58%.) The lesser destruction of volatile matter at the lower temperature when algae served as the nutrient source probably was due to the fact that the algae were able to survive under the digester conditions at 35°C; whereas they were killed at 50°C. This conjecture was confirmed in later studies in which it was found that operating the digester at 45°C led to successful digester performance.

Volume of gas produced per pound of volatile matter introduced, extent of volatile solids destruction, and the overall digester performance were not affected by the presence of aluminum in the feed. Unfortunately, at the time the experiments were conducted, it was not feasible to determine the aluminum concentration of the algal feed. Dosages of aluminum sulfate added to precipitate the algae out of suspension ranged from 90 to 120 mg/l. Digester performance remained about the same at detention periods within the range of 11 to 30 days (i.e., the optimum detention period range). The top loading rate applied in the study was 0.18 pound volatile matter per cubic foot of digester volume.

In summary, it can be stated that with the temperature at 50°C, a detention period from 11 to 30 days, and a loading rate from 0.09 pound of volatile matter per cubic foot of digester volume (minimum tried) to 0.18 pound of volatile matter per cubic foot of culture volume (maximum tried), approximately 8 cubic feet of gas will be produced of which 2.5 cubic feet will be CO_2, 5.0 cubic feet methane, and 0.5 cubic foot hydrogen, nitrogen, and other gases.

Integrated Conversion System

The development of the integrated system for converting light energy to the chemical energy of methane is described in the paper "Biological Conversion of Light Energy to the Chemical Energy of Methane," which was published in *Applied Microbiology* in 1959 (3).

The Experiments: The experimental unit used in the study is diagrammed in Figure 7.1. In its operation, decomposition products from the breakdown of organic material (algae and dead bacteria) by aerobic and anaerobic bacteria, together with light, furnished the raw material for algal photosynthesis, as a result of which light energy was fixed into algal cells. Dead algal cells, in turn, served as nutrient for the bacteria. As a result the elements of organic matter were repeatedly broken down and resynthesized. The daily procedure in operating the sections of the unit involved the following steps.

Step 1: An aliquot of algal suspension was withdrawn from the algal growth unit and was centrifuged. The supernatant was decanted and reserved.

Step 2: An aliquot was then removed from the digester.

Step 3: The algal solids were added to the digester, after they had been diluted with a part of the reserved algal culture supernatant to a volume equal to that of the aliquot removed from the digester in Step 2.

Step 4: A sample, equivalent in volume to that of the aliquot from the digester, was removed from the activated sludge unit.

Step 5: The aliquot from the digester was added to the activated sludge unit.

Step 6: The sample obtained from the activated sludge unit was diluted with the remainder of the supernatant of the aliquot from the algal culture, and was then added to the algal growth unit.

The size of the aliquots removed from each of the units was determined by the length of the detention period at which the unit was operating. The activated sludge unit was discontinued on the 110th day, and during the remainder of the study the digesting sludge was added directly to the growth unit together with the supernatant obtained from the growth unit. The entire exchange of materials was accomplished in all units with the admission of very little outside gas by simultaneously opening both the inlet and the sampling ports so that as the feed entered the top of the unit, an equivalent amount was being withdrawn from the bottom of the unit.

Gas pressures in the units were affected by the removal and admission of liquids in such a manner that no direct contact with the outside atmosphere was necessary.

The algal growth unit was operated on a detention period of 4 days for the first 20 days; of 5 days until the 70th day; and of 6 days thereafter. The activated sludge unit and the digester were operated on a 20-day detention period. The temperature of the algal culture was maintained at 26° to 27°C; of the digester cul-

FIGURE 7.1: SCHEMATIC DRAWING OF THE INTEGRATED ENERGY CONVERSION SYSTEM

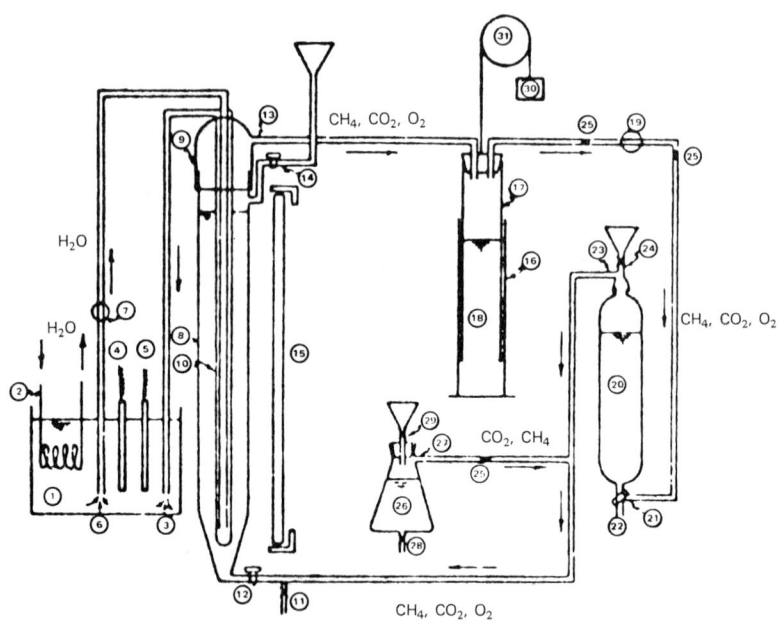

1 Water reservoir	9 Glass dome with ground glass joint	16 Gas collector	25 One-way flutter valve
2 Copper coil—tap water		17 Telescoping tube	26 Digester
3 Effluent-cooling tube	10 Cooling tube	18 Mercury	27 Gas exit
4 Heating element	11 Sampling port	19 Pump-gas recycling	28 Sampling port
5 Thermo switch	12 Gas inlet	20 Activated sludge unit	29 Feeding port
6 Influent-cooling tube	13 Gas exit	21 Gas inlet	30 Counterbalance
7 Recirculating pump	14 Feeding port	22 Sampling	31 Pulley
8 Growth unit tube	15 One of 9 fluorescent lamps	23 Gas exit	
		24 Feeding port	

Source: PB 231 149

ture, 45°C; of the activated sludge unit, room temperature. Selection of the temperature of 45°C at which to maintain the digester culture was based on the fact that it was the minimum temperature level at which the algal cells would be killed. A higher temperature was not chosen because of the increased energy required to maintain the higher temperatures. Moreover, previous work on the temperature effects on the anaerobic digestion of raw sewage sludge had shown that in a practical situation the small gain in digester performance brought about by maintenance at a thermophilic level (i.e., 50°C and higher) does not warrant the added expenditure of energy and money in a practical situation.

Light intensity at the face of the growth unit varied from 1,820 foot candles at the brightest point to 450 foot candles at the darkest point. The total amount of visible light energy available to the culture was 121,000 calories per day. Except for a slight drop in gas production over a period lasting from the 90th to the 100th day of operation, performance of the system was quite uniform until around the 190th day. For an undetermined reason, the digester culture began to fail shortly thereafter, and by the 210th day the system ceased functioning.

Gas production by the digester averaged about 10 cubic feet per pound of volatile matter introduced. From 68 to 74% of the gas was in the form of methane. Maximum conversion efficiency attained by the algal culture was 3%, whereas the maximum overall efficiency of the entire conversion unit (light energy to the energy of methane) was about 2%.

An estimate was made of the economics of a system based on a 2,000-acre algae production pond, and the pond-digester system having a net capacity of 15 kilowatts per acre (4). The estimated cost proved to be 16 mils per kilowatt hour. The cost did not take into consideration monetary savings (10 to 20%) that would accrue from the sewage treatment (secondary and tertiary) accomplished as a part of the process.

With respect to the search for new energy sources, the significance of the results of the study is that they showed that light energy can be converted to the chemical energy of methane by biological means within a relatively brief time with equipment and knowledge already available. The fact that the system could operate successfully as a closed system except for energy and the loss of some carbon in the form of gas for an appreciable length of time, means that the conversion of solar energy into methane by algal and bacterial cultures need not be restricted by limitations in quantity of organic wastes available at any one time.

The reason is that each day's contribution of nitrogen, phosphorus, and other elements is retained within the system and is used repeatedly. By recycling digested algae from the digester to the algal culture, in which the digester organic matter is further broken down by aerobic bacteria, and the resultant breakdown products serve as nutrient for the algae, digested sludge is upgraded in energy content as a result of the photosynthetic activity of the algae, and thus new organic matter becomes available and a new cycle is begun.

By the same token, the system can be strategically located both with respect to optimum cultural conditions for the algae and the power needs of the country. The portents for the economic feasibility of such a system seem to be quite favorable.

UNIVERSITY OF PENNSYLVANIA AND UARL STUDIES

This report was a result of combined efforts at the University of Pennsylvania and the United Aircraft Research Laboratories (UARL).

At the University of Pennsylvania, the goals of the study were the optimization of methane production and the achievement of an understanding of the underlying microbiological principles which regulate this system. This was undertaken by means of laboratory studies of single stage and two stage digestion.

The research effort at the United Aircraft Research Laboratories was focused in several directions. Assessment of the impact of the rapid increase in the rate of consumption of fossil fuels, both national and world wide, on time to depletion of such fuels and on the increase in global temperature and pollution prompted United Aircraft Research Laboratories in 1968 to initiate a study of means to provide a large and continuing supply of clean, easily storable fuel by conversion of solar energy to plant tissue and the conversion of the plants produced by photosynthesis to a clean, gaseous fuel (methane). Use of fuel produced in this way would result in no change in global heat balance and in no build up, on a global scale, of carbon dioxide in the atmosphere.

Results of this study indicated that under conditions existing in 1969, harvesting and conversion of crops grown specifically for the purpose of fuel-production was not economically viable. Waste materials resulting from prior use of plant crops could however, in selected cases, provide a supply of organic material for biological conversion to methane at a cost consistent with that of natural gas.

A program, sponsored by Columbia Gas Systems Service Corporation, was undertaken in 1970 to further evaluate the production of methane by conversion of animal wastes and domestic solid wastes and to carry out a systems study of the process. This included collection, transportation, waste pretreatment, biological conversion, residue disposal, gas purification and compression.

This program demonstrated a satisfactory conversion of organic material to methane and reaffirmed the estimation that, for selected areas of large population, the conversion of wastes to methane could produce a fuel competitive in price with pipeline gas at that location. It also clearly demonstrated that conversion of animal wastes and domestic solid wastes to methane, although producing a large amount of fuel, could not supply a large fraction of the national requirement for natural gas.

Further work was carried out at United Aircraft Research Laboratories directed to the evaluation of harvested plants for conversion to methane gas. Both land plants and water grown plants were considered. The area needed for plant growth equivalent to 1971 natural gas requirements in the United States was estimated to be of the order of 10^5 square miles. Although this is not a large fraction of the U.S. land/water area, it was apparent that serious consideration should be given to finding the required area for photosynthesis in selected parts of the oceans. One configuration of this concept required conversion of the plant crop to methane in biological conversion units adapted to a marine environment and development of such seawater adapted bacterial cultures was achieved.

The joint program involving United Aircraft Research Laboratories and University of Pennsylvania had as its objectives the continued evaluation of the technical and economic feasibility of the conversion of organic materials to methane by biological fermentation under anaerobic conditions. The organic materials of interest included animal wastes and domestic solid wastes, harvested land plants (grasses), fresh water plants (water hyacinth and unicellular algae), marine plants (seaweed and giant kelp). These were to be studied individually and in combinations in marine environment and in the conventional fresh water environment.

Materials and Methods

This section of the report is divided in two sections, thus separating the UARL and University of Pennsylvania (UP) studies.

UARL Laboratory Digesters: Digesters are of cylindrical shape and operate with liquid volumes of 10 liters. They are constructed to provide means for addition of feed materials, removal of liquid and solid wastes, stirring of the liquid contents, and removal of the gas produced. The digesters are heated to the desired temperature either by individual heating cables attached to the outside of the digesters or by placement of the digesters in a large temperature controlled cabinet. Temperatures are controlled to ± 0.5°C. The digesters are fabricated from aluminum or Plexiglas.

The contents of aluminum digesters are stirred by means of a gas lift pump. The gas above the level of liquid in the digester is compressed by means of a small diaphragm pump and discharged through an aluminum tube near the base of an aluminum chimney mounted in the liquid. The liquid within the chimney is raised and spilled over the top by the gas introduced. Bottom liquid enters the chimney and circulation is established. The stirring of the contents of the Plexiglas digesters during earlier studies had been by means of centrifugal pumps. This procedure has not been used during this study and the only stirring is provided by gas generation. Based on digester performance this has been adequate.

The coarseness of the feed materials used in this study has precluded automatic feeding on a continuous or semicontinuous basis. Thus, the feed materials are added on a once daily batch basis. The feed may be in the form of a slurry of known concentration or it may be of the coarsely ground material with no added water. Amounts of digester contents are removed on a daily basis to retain the desired volume of slurry within the digester. The amount and composition of the gas produced within each digester is monitored daily.

UARL Digester Operation: Start-Up — The digester inocula have consisted of effluent obtained from municipal primary sewage sludge digesters, samples of garden soil, black mud from the bottom of a fresh water pond, and the rumen contents of slaughtered cattle. Mud from a brackish-water pond on Cape Cod has been used for the initial conversion of digesters from fresh water to seawater operation. The value of the brackish water inoculum is unknown.

As a general procedure, the seed materials were placed in empty digester vessels which had been purged of oxygen by a stream of nitrogen or carbon dioxide gas. The digester vessels were maintained at the temperature to which the seed materials had been acclimated. Nutrients were supplied to the seed material in the form of a slurry of the material to be studied or of dog food. When evidence of gas production was seen, the volume of liquid in the digester was slowly increased by daily addition of the feed slurry. The liquid volumes added were usually 10% of the volume already in the digester and the nutrient addition usually amounted to about one or two grams dry solids for each liter of liquid volume in the digester.

Where dog food was used as starting nutrient, feeding was continued until stable operation was achieved. Change over to the selected feed and operating conditions was then carried out. In many cases, new digesters were inoculated with the effluent of other digesters operating under similar conditions.

Conversion to Seawater — Conversion from fresh water to seawater environment was usually performed by slow adaption of an operating fresh water digester to increasing seawater concentration. It has generally been observed that, at some intermediate concentration of seawater, the activity of the bacterial population decreases markedly. Continued operation eventually leads to recovery, after which, the seawater concentration may again be increased.

Digester Feeding — Digesters have been fed by adding the entire days' supply at one time. The material, shredded or coarsely ground, is added as a slurry in water or is introduced without added water. Normally, daily feeding is attempted but feedings are sometimes omitted on weekends and holidays. Prior to feed addition, digester contents are removed in amounts equal to the volume of feed to be added.

UARL Operating Variables: Temperature — Laboratory digesters have been operated at temperatures of 33°C, 39°C, 41°C and 48°C. For most of the studies, cultures adapted to 48°C have been used. One digester started on a municipal sewage digester effluent has been operated at 33°C and one digester started with cattle rumen contents has been operated at 39°C.

Feed Materials — The digester feed materials which have been used during this study include:

> paper (newsprint, paper towels)
> grass
> household garbage (food wastes only)
> unicellular fresh water algae
> water hyacinth *(Eichhornia crassipes)*
> seaweed *(Ascophyllum nodosum)*
> giant kelp *(Macrocystis pyrifera)*
> cattle manure
> dry dog food (for base line data)

All feed subject to decomposition (water hyacinth, seaweed, garbage and manure) is stored under refrigeration. Long term supplies are kept frozen in a commercial establishment. Grass supplies are normally air-dried and did not require refrigeration. Water hyacinth and seaweed were obtained from commercial establishments, garbage, grass and newsprint from lab personnel and cattle manure from a local farm. The garbage consists of household food wastes only and does not contain paper, glass, plastic or metal. The fresh water algae were obtained from Professor William Oswald at the University of California at Berkeley.

Water hyacinth and seaweed are pulverized in the blender after freezing with liquid nitrogen. Garbage mixed with a minimum amount of water is also homogenized in the blender. Paper (newsprint) is uniformly shredded prior to use. Multicomponent feed materials are also homogenized in the blender prior to use.

The feed composition which has been designated as "domestic solid waste" (DSW) is a mixture synthesized from newsprint, garbage and grass. The composition is 66% paper, 17% garbage and 17% grass (all on a dry basis). Purina dog chow has been used as a base line feed, and, in several cases, for digester start-up.

Identification of the seaweed was made by Professor P.E. Hargraves, Graduate School of Oceanography, University of Rhode Island, Kingston, Rhode Island.

The seaweed is the common intertidal brown algae, *Ascophyllum nodosum.*

UARL Pretreatment of Feed Materials: An observed poor performance of digesters fed with water hyacinth and seaweed suggested the possibility that a treatment to hydrolyze or otherwise destroy a resistant surface of the plants could improve operations. It was assumed that lack of a sufficient quantity of an enzyme able to hydrolyze cellulose constituted the basic problem.

A few preliminary experiments were carried out to demonstrate that addition of cellulase (the cellulose hydrolyzing enzyme) would produce initial attack on cellulose rich plants and permit further bacterial attack to occur. Effluent from one digester was placed in each of three 1-liter flasks with provisions for measurement of gases evolved. Temperature of the flasks was maintained at 48°C.

Into one flask was placed 810 ml of effluent plus 90 ml of a slurry of water hyacinth, a material found difficult to digest. In a second flask was placed the same materials plus about 2.5 grams of a crude cellulase preparation. In the third was placed 810 ml of effluent and 90 ml of a slurry of water hyacinth which had been in contact with 2.5 grams of cellulase for 24 hours.

Gas production from the three systems was monitored for 24 hours. During that period no gas was produced in the flask without cellulase; 390 ml of gas was produced in the flask containing freshly added cellulase and 330 ml of gas was produced by the flask containing the cellulase which had been in contact with the hyacinth for 24 hours prior to placing into the digester effluent.

These results indicate that digestion of water hyacinth can occur after pretreatment with cellulase. Since the use of cellulase would be prohibitively expensive, another series of tests was carried out in which hyacinth pretreated with alkali was compared with hyacinth treated with cellulase. Using the same type of test apparatus it was found that treatment of 90 ml of hyacinth slurry by boiling for 15 min in the presence of 0.25 g NaOH per 100 ml of feed and then neutralization with HCl and cooling the feed yielded an amount of gas equal to or greater than that from the cellulase treatment. Consequently, several digesters were operated during the last two months of the program with caustic pretreated feeds.

UARL Chemical Analyses: The composition of the gases produced in the digesters is established daily before addition of new feed. Analysis of methane and carbon dioxide is carried out by gas chromatography. In certain cases more complete analyses are carried out using a mass spectrometer. Samples of digester contents removed just prior to feed addition are subjected to extensive chemical analysis and characterization. The pH and alkalinity, volatile acids, total solids and volatile solids, total carbon, total nitrogen and ammonia nitrogen are determined. Volatile acids are determined by the extraction-titration method; total nitrogen and ammonia nitrogen by standard Kjeldahl procedure or by the use of a specific ammonia sensing electrode. All other analyses are conventional.

Analyses of feed materials are carried out on new lots of the various materials. The seawater used is analyzed for total salt content. The seawater is collected in hundred gallon batches from Long Island Sound as close to high tide and as far from fresh water estuaries as possible to minimize dilution with fresh water. Feed analysis and seawater data are shown in Table 7.1.

TABLE 7.1: FEED AND SEAWATER ANALYSIS

Material	Batch Number	Date	Moisture (%)	Total Solids (%)	Volatile Solids (%)	Total Carbon (%)	Total Nitrogen (%)	COD (g/g)	Ash (%)
Water hyacinth	1	9/17/73	30.4	69.9	90.78	–	–	–	9.22
	1	10/5/73	–	–	–	29.1	1.39	1.2	–
Animal waste	1	8/27/73	84.8	15.2	87.5	46.7	3.19	0.73	12.5
	2	9/24/73	87	13	87.1	–	–	–	12.9
		10/15/73	93.6	6.4	85.0	38.4	7.43	–	15
		10/18/73	83.4	16.6	86.1	42.4	5.46	1.51	13.9
		12/11/73	85.3	14.7	85.1	40.7	1.96	1.43	14.9
Paper (newsprint)	1	8/27/73	5	95	98.6	46.3	0.84	1.29	1.4
	1	8/28/73	5.3	94.7	98.6	–	–	–	1.4
Seaweed (Ascophyllum nodosum)	1*	6/21/73	9.9	90.1	80	46	3.4	0.75	20
	*	6/21/73	9.94	90.06	77.55	–	–	–	22.45
	2*	7/30/73	13.2	86.8	75.7	–	–	–	24.3
	3*	8/27/73	12.1	87.9	81.4	33.8	1.4	0.88	18.6
	Wet	8/27/73	66.6	33.7	–	–	–	–	–
Giant kelp (Macrocystis pyrifera)	Blade**	11/20/73	87.7	12.3	53.0	14.5	–	–	47
	Stem & float**	11/20/73	90.5	9.5	47.9	–	2.36	–	52.1
	Wet††	12/6/73					1.72		
	Blade††	12/13/73							–
Domestic solid waste	1	6/18/73	73.4	26.6	–	–	–	–	3
	2	6/18/73	78.9	21.1	93.4	–	–	–	6.6
	2	6/21/73	79.1	21	93.4	43.8	2.38	1.05	6.6
	3	7/30/73	73.3	26.7	94	–	–	–	6.0
	4	8/27/73	79.3	20.7	95.6	35.6	3.63	1.57	4.4
	5	8/27/73	80.7	19.3	93.7	24	4.43	1.58	6.3
	6	10/1/73	79.8	20.2	93.1	–	–	–	6.9
	6	10/5/73	–	–	–	49.2	6.9	1.79	–

(continued)

TABLE 7.1: (continued)

Material	Batch Number	Date	Moisture (%)	Total Solids (%)	Volatile Solids (%)	Total Carbon (%)	Total Nitrogen (%)	COD (g/g)	Ash (%)
Grass	1	6/18/73	9.95	90.05	91.1	–	–	–	8.9
	2	6/21/73	16	84	85	34	3.9	0.85	15
	3	8/27/73	7	93	91.2	46.1	3.6	1.11	8.8
Algae	Bag #1	9/20/73	8.1	91.9	–	–	–		–
	Bag #2	9/20/73	8.35	91.65	–	–	–		–
	Bag #1	9/24/73	8.3	91.7	92.6	–	–		7.4
	Bag #2	9/24/73	8.5	91.5	92.8	–	–		7.2
	Bag #2	10/18/73	7.7	92.3	93.0	43.0	9.1		
Seawater	1	5/7/73		3.02					
	2	6/7/73		3.26					
	3	8/27/73		3.17					
		9/11/73		3.34					
Purina	1	4/3/73	6.3	93.7	91	38.8	3.75	0.71	9
Dog	2	6/21/73	6.5	93.5	90.1	36.6	3.36	0.86	9.9
Chow	3	8/26/73	6.3	93.7	90.6	37.4	3.12	1.19	9.4
		8/26/73	6.4	93.6	90.4	39.7	3.47	1.24	9.6

*Air dried **As received †With seawater ††Plus stem & float

Note: All calculations based on samples dried at 103°C.

Source: PB 238 103

UP Analytical: The analytical back-up for the laboratory scale digester studies consists of standard wet chemistry methods for the determination of digester operating parameters such as ammonia nitrogen, organic nitrogen, total Kjeldahl nitrogen, total solids, volatile or organic solids (VS), inorganic solids, chemical oxygen demand (COD), biochemical oxygen demand (BOD), alkalinity, etc. Additionally, the capability exists for the determination of calcium and ammonia. These analyses are performed to monitor digester feed, supernatant, digested solids and digester performance.

Gas chromatography (Beckman GC-72-5) is used for the determination of volatile acids composition. The gas chromatograph is equipped with thermal conductivity and flame ionization detectors.

A three column system is used for gas analysis. The silica gel column adsorbs CO_2; the molecular sieve column separates O_2, H_2, N_2 and CH_4; and the Porpak-T separates H_2S and CO_2. A thermal conductivity detector is used for gas analysis.

The analysis of volatile fatty acids is performed using a column packed with 10% SP-1200 and 1% H_3PO_4 on AW Chromosorb W 80/100 mesh. The column is located in place of the Porpak-T column. Volatile acids are determined with the flame ionization detector. The capability exists to quantify the following potential methanogen substrates: formic, acetic, propionic, n-butyric, isobutyric, n-valeric and isovaleric acids.

The procedure for extracting the volatile acids prior to injection into the gas chromatograph consists of the following steps: (1) acidification of a known sample volume to pH 1.5 using concentrated H_2SO_4; (2) addition of a known volume of ether for extraction; and (3) removal of an aliquot of the ether phase for injection.

UP Digesters: The laboratory-scale fermentors have been designed and constructed to yield maximum flexibility, automatic operation, and automatic and continuous recording of all pertinent data. The fermentation digestion equipment is housed in a specially constructed cabinet. The cabinets are divided into three sections. One side, the "wet section," houses digesters, while the other side, the "electronics section," contains the automatic measuring and control equipment and recording instruments. In addition, the lower part of the cabinet consists of a catch basin of sufficient volume to hold the total liquid contents of the digesters in the event of an accident.

The wet side has sufficient room for two digesters of at least 20 liters capacity each. This side is completely enclosed in Plexiglas for easy observation and effective containment and venting. Two adjacent sides of this enclosure are hinged doors to present easy and full access to all equipment. The fermentation vessels are equipped with sensing probes for the measurement and control of pH and temperature. Constant temperature is maintained by a heater element submerged below the level of the digester contents. Mixing is provided by variable speed propellors.

A valve located at the bottom of the digester allows sample collection and drainage of the digester contents or of settled digested solids. The digesters are cylindrical Plexiglas vessels. Access of the atmosphere is prevented by means of an O-ring seal between the vessel and its top plate, and of a mercury seal between the top plate and the propellor shaft. The digester lid contains a series of ports for the entry of the heater element, sensing probes and various gas and liquid lines. The latter include gas exit, N_2 entrance (for flushing the digester to maintain anaerobiosis during those periods when the digester is opened for maintenance), feed entrance and pH control (Figure 7.2).

FIGURE 7.2: SCHEMATIC OF DIGESTER TANKS USED FOR SINGLE STAGE AND TWO STAGE DIGESTION

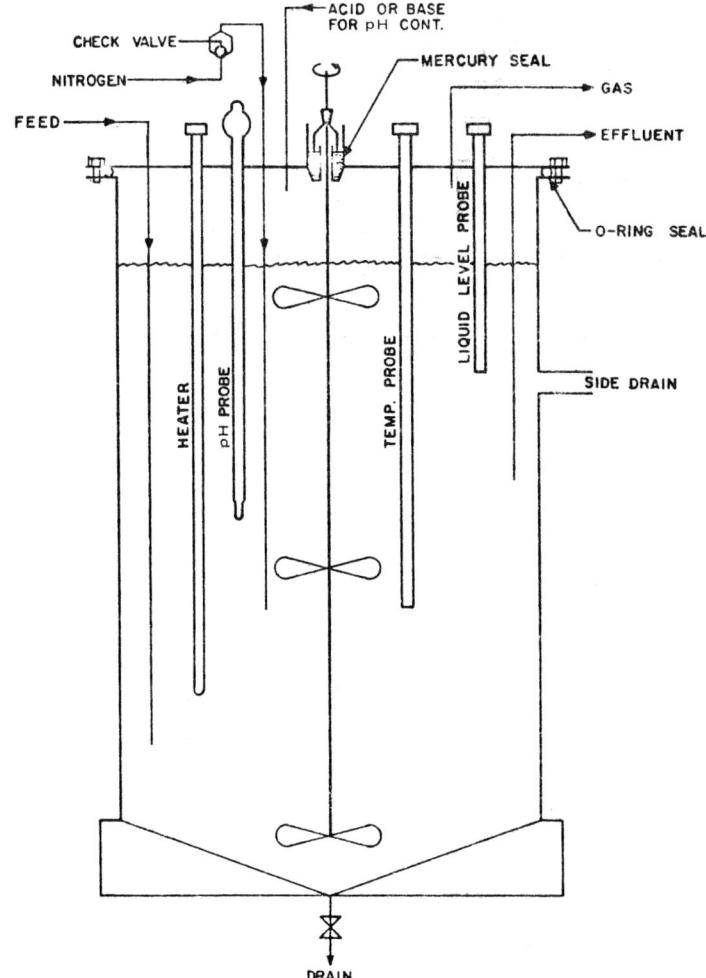

Source: PB 238 103

186 Energy from Bioconversion of Waste Materials

The electronics section of the cabinet houses the required instrumentation and control. Wet test meters are used to measure the volume of gas produced. The measured volumes are then converted to standard conditions. pH meters receive the signal from the pH probe within the digesters. The pH meters in turn activate the pumps which add acid or base to the digesters in order to maintain the desired H^+ concentration. Any data taken on a single, or two-stage digester system are continuously recorded on a single chart by a 12 channel recorder. (This facilitates examination of the data.) Such systems are used to continuously monitor pH, temperature and gas production volume.

The digestion system used offers many advantages over conventional laboratory-scale digesters; in particular, continuous and automatic control over all operational procedures, including feeding at regular intervals with controlled amounts, and pH control. These facilities ensure successful operation over holiday and weekend periods. The continuous recording of digester status (parameters such as pH, temperature, redox potential, gas production, etc.) provides a thorough data base.

The digesters are either operated as single stage digesters, in which both the acidogenic and methanogenic bacteria are present in a single tank, or, as two stage digesters, in which the groups of bacteria are separated into two tanks operating in series. A schematic of the two-stage operation appears in Figure 7.3.

FIGURE 7.3: SCHEMATIC OF TWO-STAGE DIGESTION SYSTEM

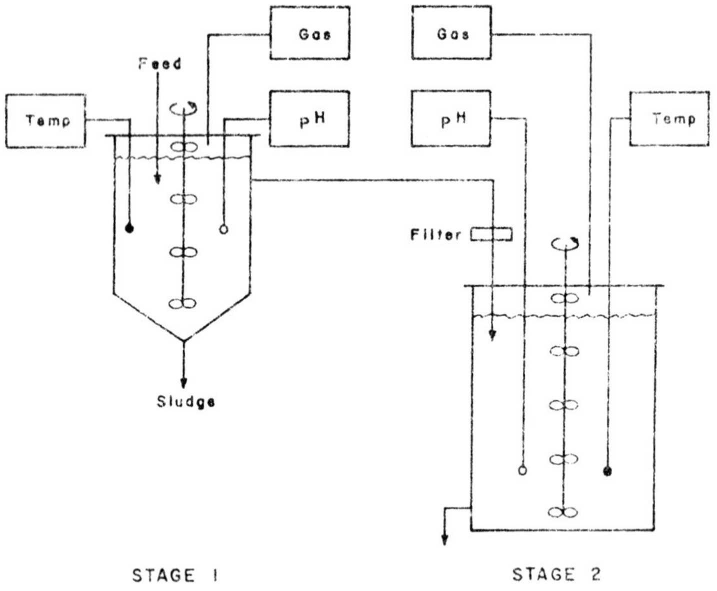

Source: PB 238 103

UP Feeding Procedures: The digesters are fed on a once daily (seven days/week) batch basis. Dried dog food is used as feed material. The digesters are normally loaded at 5 g dry organic matter/liter of culture volume/day (0.31 lb VS/ft^3/day). The feed material is added as a 5% slurry, after draining an equivalent volume of effluent. Small particle size in the feed is produced by soaking in water, followed by blending (Waring Blender) for two minutes.

The strict redox potential requirements of the methanogens are met by preventing access of oxygen to the culture. This is done by boiling the dilution water, flushing the feed thoroughly with N_2 gas and by flushing the gas space over the culture with N_2. The second stage digester is fed by direct transfer of Stage 1 effluent to the Stage 2 tank.

Results of Single-Stage Digestion—UARL

At the start of the period, fourteen digesters were in operation. As a result of a decision reached with University of Pennsylvania to emphasize the study of the use of the giant kelp, *Macrocystis pyrifera,* as a material for cultivation and conversion to fuel gas, ten of the digesters were deactivated. Those deactivated included all digesters in which the feed contained paper, garbage, grass or algae.

Also included were one of two digesters utilizing common seaweed (*Ascophyllum nodosum*), one of two digesters in which the feed was water hyacinth (*Eichhornia crassipes*), and one seawater digester maintained as a source of seed material. The digesters remaining in operation included one utilizing animal waste, one with seaweed, one with water hyacinth and one which was in the process of conversion from fresh water to seawater operation. Deactivation of the digesters involved only the cessation of feeding and analysis of digester liquid and gas. Temperature control and recording of gas production was continued.

In order to study the conversion of Macrocystis to methane, two new fresh water digesters were started, one with cattle rumen contents and one with effluent from a municipal sewage sludge digester operating at 33°C. Later, one of the seawater digesters was reactivated and also used with Macrocystis as the only feed.

These changes resulted in the active operation of six digesters with feed materials of potential interest for fuel gas generation, and of one digester to provide further information on acclimation to seawater environment.

Operating results are presented first for those digesters which were actively maintained or started during the latter period. Results obtained with the other digesters prior to their deactivation are then presented.

Conversion of Seaweed in Fresh Water Digesters: To explore the possibility of converting sea grown plants to fuel gas in fresh water digesters, two digesters were started. One digester, designated 8A, with a liquid volume of 28 liters, was started with effluent from a municipal sewage sludge digester which had been operating at 33°C. This temperature was maintained in the laboratory digester.

A second digester, designated 7A, with a liquid volume of 28 liters, was started with contents of cattle rumen. This digester was operated at 39°C. The feed to both of these digesters was the common seaweed (*Ascophyllum nodosum*) which was added to the digesters, after grinding, as a slurry in water at a loading of 2.0 g total solids/liter/day. The hydraulic detention time was 20 days. This was accomplished by adding 1.4 liters of a 4% slurry to the digesters each day. To maintain constant volume in the digester, 1.4 liters of digester contents were removed each day. Data for each digester are presented below.

Digester 8A — Feeding with seaweed at 2 g/l/day was started on October 25, 1973. Gas production was essentially zero during a period of about 3 weeks. This lack of activity could have been due to the inability of the digester to adapt to the new feed material or to an inherent property of the seaweed preventing rapid attack by the bacterial population present in the digester. On the basis of analyses of the digester contents, it was apparent that there was not a high concentration of volatile acids developing, and that other parameters were acceptable. Therefore, it was concluded that there was no attack on the seaweed particles.

A procedure was initiated in which the seaweed was heated in the presence of a small quantity of alkali, cooled and neutralized with acid and then added to the digester. This procedure was utilized in order to initiate hydrolysis of the feed material.

Gas production soon became significant and continued at a moderate level. The gas contained about 50% CH_4 and its production rate was 4 to 5 liters (STP)/day. During a 3-week period, 84 g dry solids in the form of seaweed were added to the digester. The composition of the material varied with the amount of occluded salts. However, this amount of seaweed represented about 300 g carbon. The total methane production during the test period was about 46 liters (STP) containing about 25 g of carbon. The conversion efficiency of seaweed to methane in this digester is therefore 8.3%. The data relating to operation of this digester with seaweed are presented in Tables 7.2 and 7.3.

Digester 7A — Feeding with seaweed (*Ascophyllum nodosum*) started at the loading and detention time indicated above. No significant gas production occurred during a period of about three weeks. The pH of this digester was quite low and appropriate amounts of lime were added to bring the pH up to a more acceptable level. Volatile acids in this system were quite high during this period. Other parameters were acceptable.

After three weeks, feeding of the digester with alkali treated seaweed was started. Increase in gas production was immediate and the gas analysis indicated more than 50% methane. There was also a rapid drop in volatile acids to a value about $1/6$ of that observed with untreated seaweed. The gas production rate increased to about 18 liters (STP)/day and then decreased again to about 10 liters (STP)/day.

During the test period (11/16 to 12/7/73), 780 g of seaweed was added to the digester. This corresponds to 290 g carbon. The total gas production during this time amounted to 235 liters (STP) containing 55% CH_4. Seventy grams of carbon are contained in the methane produced and the conversion efficiency based on carbon is 24%. The data obtained for this period of operation of digester 7A are presented in Tables 7.4 and 7.5.

Conversion of Macrocystis in Fresh Water Environment: The conversion to

TABLE 7.2: DIGESTER 8A–CONVERSION OF SEAWEED TO METHANE FEED (UNTREATED SEAWEED)

Date	Gas Production Rate (l, STP/day)	Gas Production Composition CH_4/CO_2	pH	Alkalinity (mg $CaCO_3$/l)	Volatile Acids (mg/l)	Total Solids	Volatile Solids	Carbon	Total Nitrogen	NH_3-N (mg/l)
10/26/73	0.2		6.88		132					
10/27/73	0									
10/28/73	0									
10/29/73	0	1.30	6.78	2,200		20,670	14,500	7,420	1,380	293
10/30/73	2.6	1.46	6.90							
10/31/73	0.2	1.28	6.99							
11/1/73	0	1.30	6.94	2,100		16,140	11,650	6,120	915	275
11/2/73	0	1.35	6.92							
11/3/73										
11/4/73	0	1.38								
11/5/73					84					
11/6/73	7.4	1.33	6.83	1,900		12,800	9,140	4,660	720	282
11/7/73	1.6	1.28	6.86							
11/8/73	0	1.26	6.92	1,900	60	11,650	8,350	3,590	660	250
11/9/73	0	1.22	6.71		48					
11/10/73										
11/11/73	0.9	1.33								
11/12/73			6.90							
11/13/73	0	1.17	6.80	2,000		11,000	7,840	2,300	490	210
11/14/73	0	1.17	6.85							
11/15/73	0	1.16	6.68							
11/16/73	0	1.17	6.87		72					

NOTE: Loading, 2.0 grams total solids/liter/day; detention time, 20 days; temperature, 33°C.

Source: PB 238 103

TABLE 7.3: DIGESTER 8A—CONVERSION OF SEAWEED TO METHANE FEED (CAUSTIC TREATED SEAWEED)

Date	Gas Production Rate (l, STP/day)	Composition CH_4/CO_2	pH	Alkalinity (mg $CaCO_3$/l)	Volatile Acids (mg/l)	Total Solids	Volatile Solids	Carbon	Total Nitrogen	NH_3-N (mg/l)
11/17/73										
11/18/73	0.3	1.23								
11/19/73			6.95	1,700		9,200	6,070	2,510	400	172
11/20/73	0	0.80	6.85							
11/21/73	3.0	0.82	6.85							
11/22/73										
11/23/73	1.5									
11/24/73			6.68							
11/25/73	0.4	1.15								
11/26/73			6.61		72					
11/27/73	6.7	1.05	7.02	2,400		10,000	5,040	1,870	390	166
11/28/73	7.5	1.01	6.87							
11/29/73	6.7	1.00	6.88		672					
11/30/73	5.8	1.00	6.80	2,900		13,540	6,380	2,900	275	127
12/1/73										
12/2/73	14.7	1.20								
12/3/73			6.98		1,224					
12/4/73	4.2	1.14	7.04	2,800		15,230	7,000	3,060	295	133
12/5/73	4.0	1.13	7.02							
12/6/73	5.3	0.79	6.98							
12/7/73	3.2	0.96	7.09	3,000	1,752	19,840	8,840	4,820	450	135

NOTE: Loading, 2.0 grams total solids/liter/day; detention time, 20 days; temperature, 33°C.

Source: PB 238 103

TABLE 7.4: DIGESTER 7A–CONVERSION OF SEAWEED TO METHANE FEED (UNTREATED SEAWEED)

Date	Gas Production Rate (l, STP/day)	Gas Production Composition CH₄/CO₂	pH	Alkalinity (mg CaCO₃/l)	Volatile Acids (mg/l)	Total Solids	Volatile Solids	Carbon	Total Nitrogen	NH₃-N (mg/l)
10/26/73	24.0	–	6.0		3,910					
10/27/73										
10/28/73	10.0	0.46		3,050		10,370	5,880	2,880	470	188
10/29/73			5.70							
10/30/73	1.7	0.41	6.13							
10/31/73	0.3	0.48	6.22							
11/1/73	0.1	0.75	6.55	3,100		11,300	6,710	3,130	310	192
11/2/73	0.3	0.88	6.71							
11/3/73	0									
11/4/73	0									
11/5/73	0	–	–		5,580					
11/6/73	0	–	–	3,500		11,730	7,030	3,430	342	188
11/7/73	0.1	–	–							
11/8/73	0.2	–	6.06	3,200	6,720	10,960	5,780	2,830	225	172
11/9/73	0.3	5.1	6.12		6,600					
11/10/73										
11/11/73	0.6		–							
11/12/73										
11/13/73	0.5	2.5	6.38	3,200		11,140	6,460	2,120	195	162
11/14/73	0.5	2.0	6.77							
11/15/73	0.1	2.8	6.82							

NOTE: Loading, 2.0 grams total solids/liter/day; detention time, 20 days; temperature 39°C.

Source: PB 238 103

TABLE 7.5: DIGESTER 7A–CONVERSION OF SEAWEED TO METHANE FEED (CAUSTIC TREATED SEAWEED)

Date	Gas Production Rate (l, STP/day)	Gas Composition CH_4/CO_2	pH	Alkalinity (mg $CaCO_3$/l)	Volatile Acids (mg/l)	Total Solids	Volatile Solids	Carbon	Total Nitrogen	NH_3-N (mg/l)
11/16/73	0.9	10.5	7.10		6,360					
11/17/73										
11/18/73	3.0	17.0								
11/19/73			7.30	3,500		11,260	6,300	2,740	190	175
11/20/73	2.2	1.75	6.90							
11/21/73	6.4	1.64	7.05							
11/22/73	18.9	–								
11/23/73			7.24							
11/24/73	17.3	1.42								
11/25/73					744					
11/26/73			7.02							
11/27/73	16.8	1.26	7.42	5,100		17,600	10,300	4,900	470	112
11/28/73	18.0	1.22	6.95							
11/29/73	14.0	1.20	7.31		1,044					
11/30/73	12.0	1.24	7.10	4,300		17,720	9,270	4,160	282	76
12/1/73										
12/2/73	29.7	1.22								
12/3/73			7.23		1,770					
12/4/73	9.7	1.24	7.21	4,100		19,050	9,570	4,100	273	73
12/5/73	10.2	1.18	7.18							
12/6/73	11.9	1.20	7.16	3,900	1,440	20,370	9,220	4,520	385	104
12/7/73	9.3	1.24	7.30							

NOTE: Loading, 2.0 grams total solids/liters/day; detention time, 20 days; temperature, 39°C.

Source: PB 238 103

methane of a second type of marine plant in fresh water was examined. Digesters 8A and 7A had been fed giant kelp, *Macrocystis pyrifera,* for a brief period of time.

Digester 8A was maintained at 33°C and was fed a slurry of Macrocystis at a loading of 2 g total solids/liter/day. Gas production increased from an initial rate of about 3 liters (STP)/day to about 10 liters (STP)/day. The composition of the gas was about 33% methane and 67% carbon dioxide. Thus, it is apparent that Macrocystis is subject to bacterial attack in a fresh water digester. Over the short period of the study, approximately 29 g carbon as CH_4 were produced from approximately 120 g of carbon present in the Macrocystis feed. This corresponds to a conversion efficiency of approximately 24% based on methane carbon and feed carbon.

Digester 7A, operating at 39°C, was fed Macrocystis at the same loading as Digester 8A. The gas production rate was approximately the same as with 8A and the methane content, although initially greater than 50%, decreased, during the test, to about 40%. Very few analyses of digester liquid were carried out during these tests with Macrocystis so no material balances across the systems could be made.

It should be pointed out that the salts present in the seawater and Macrocystis and in occluded brine are being added daily to a fresh water system and would, therefore, soon reach a significant level. Whether the digesters would be able to adapt rapidly to such conditions is unknown. The data for Digesters 8A and 7A operating with Macrocystis are presented in Tables 7.6 and 7.7.

Conversion of Water Hyacinth in Fresh Water Digester — Digester 7B, operating at 48°C, was used to establish the feasibility of converting the plentiful and easily harvested water hyacinth to methane. At the start of the test the digester was operating satisfactorily with all key parameters at acceptable levels.

On September 27, 1973 slurry feed of water hyacinth at a loading of 4.0 g total solids/liter/day was initiated. The liquid volume of the digester was 28 liters, and the daily volume of slurry added and effluent removed was 2.8 liters. Therefore the hydraulic residence time was 10 days. When the feed was changed to hyacinth there was an abrupt decrease in gas production rate, dropping from about 40 liters (STP)/day to about 2 liters (STP)/day in a period of less than two weeks. The build-up of solids in the digester without a significant increase in volatile acids in the system suggests that hyacinth is resistant to attack by the bacterial population.

Then the use of hyacinth pretreated with heat and caustic was started. Loading and detention time remained the same. There was an immediate increase in gas production rate to about 8 to 9 liters (STP)/day but this rate fell to about 2.5 to 4.0 in a few days and remained at this level for the duration of the experiment. There was a large build-up of solids in the digester, some of which was due to the ash content of the plants. Data for this test are presented in Table 7.8.

Conversion of Macrocystis in Seawater Environment — A desirable environment for the conversion of Macrocystis to fuel gas would be one identical to that in which the plants had been grown. Consequently, two seawater digesters

TABLE 7.6: DIGESTER 8A—CONVERSION OF KELP TO METHANE

Date	Gas Production Rate (l, STP/day)	Gas Production Composition CH₄/CO₂	pH	Alkalinity (mg CaCO₃/l)	Volatile Acids (mg/l)	Total Solids	Volatile Solids	Carbon	Total Nitrogen	NH₃-N (mg/l)
12/8/73										
12/9/73	9.3	0.95								
12/10/73	3.0	1.00	7.12	3,000		20,430	10,230	4,900	341	133
12/11/73	2.9	0.99	7.04		1,464					
12/12/73	3.5	0.90	7.02							
12/13/73	3.8	0.81	7.10	3,200		22,350	10,920		360	116
12/14/73			7.07		528					
12/15/73										
12/16/73	14.7	0.68								
12/17/73			6.88	3,700		44,080	26,370		1,110	107
12/18/73	8.2	0.51	6.82		1,848					
12/19/73	9.4	0.44	6.90							
12/20/73	9.2	0.43	6.95							
12/21/73	9.9	0.40	6.95							
12/22/73										
12/23/73	20.0	0.49	6.90							
12/24/73										
12/25/73	10.7	0.67	7.00							
12/26/73										
12/27/73	13.4	0.70	7.10							
12/28/73										
12/29/73	8.9	0.69	7.01							
12/30/73										
12/31/73	11.3	0.69	6.90							

NOTE: Loading, 2.0 grams total solids/liter/day; detention time, 20 days; temperature, 30°C.

Source: PB 238 103

TABLE 7.7: DIGESTER 7A–CONVERSION OF GIANT KELP TO METHANE FEED (MACROCYSTIS)

Date	Gas Production Rate (l, STP/day)	Gas Production Composition CH$_4$/CO$_2$	pH	Alkalinity (mg CaCO$_3$/l)	Volatile Acids (mg/l)	Total Solids	Volatile Solids	Carbon	Total Nitrogen	NH$_3$-N (mg/l)
12/8/73										
12/9/73	23.4	1.22								
12/10/73	8.2	1.26	7.38	3,900		19,420	9,230	4,330	270	85
12/11/73	7.1	1.30	7.52		1,470					
12/12/73	8.4	1.13	7.25							
12/13/73	9.4	1.00	7.18	3,800	1,020	20,300	8,304		255	66
12/14/73			7.20							
12/15/73										
12/16/73	23.4	1.13								
12/17/73	6.5	1.06	7.20	4,100		22,640	9,860		405	85
12/18/73	6.8	1.00	7.16		1,530					
12/19/73	8.4	1.00	7.15							
12/20/73	9.4	0.92	7.20							
12/21/73			7.25							
12/22/73										
12/23/73	16.2	1.17	7.15							
12/24/73										
12/25/73	7.0	1.24	7.25							
12/26/73										
12/27/73	9.6	0.91	7.15							
12/28/73										
12/29/73	9.2	0.88	7.17							
12/30/73										
12/31/73	11.4	0.77	7.09							

NOTE: Loading, 2.0 grams total solids/liter/day; detention time, 20 days; temperature, 39°C.

Source: PB 238 103

TABLE 7.8: DIGESTER 7B—CONVERSION OF WATER HYACINTH TO METHANE

Week No.	Gas Production Rate (l, STP/day)	Composition CH_4/CO_2	pH	Alkalinity (mg $CaCO_3$/l)	Volatile Acids (mg/l)	Total Solids	Volatile Solids	Carbon	Total Nitrogen	NH_3-N (mg/l)
\multicolumn{11}{c}{Untreated Water Hyacinth}									
1	12.4	1.8	7.15	2,220	120					
2	2.4	1.7	6.90	1,340	150	2,190	1,580	705	375	265
3	1.5	1.5	6.70	1,620	360	8,400	6,600	3,010	320	105
4	4.3	2.0	7.40	3,000	396					
5	2.7	2.0	7.60	2,250	240	5,400	4,120	2,210	780	442
6	1.8	2.0	7.20	1,750	72	5,290	4,060	2,600	835	328
7	1.5	1.8	6.80	1,300	156	2,680	1,830	930	246	225
\multicolumn{11}{c}{Caustic Treated Hyacinth}									
1	6.5	1.3	7.0		240					
2	3.4	1.3	7.05	1,770	54	17,950	8,930	4,430	350	200
3	2.9	1.2	6.95	1,300	96	15,320	4,380	2,030	300	170
4	3.1	1.15	7.00	1,150	102	16,250	4,070	–	265	142
5	2.9	1.2	7.00		84					
6	1.8	1.2	7.10							

NOTE: Loading, 4.0 grams total solids/liter/day; detention time, 10 days; temperature, 48°C.

Source: PB 238 103

(1B and 8B) were fed Macrocystis. The loading of Digester 1B was 2 g total solids/l/day with a hydraulic detention time of 20 days. Digester 8B was loaded at the rate of 1 g total solids/l/day and the detention time was 50 days for 10 days of the test and then at an effective loading of 0.5 g/l/day for an additional 12 days.

Conversion of Macrocystis to gas in Digester 1B was only 7%. This digester had been operating poorly for several months. The relatively low pH values and the high CO_2 content of the gas produced indicated that the bacterial population was deficient in methane producers.

In Digester 8B, gas production was quite satisfactory, and, based on the carbon content of the feed and an estimate of the fraction of total gas ascribable to Macrocystis, the efficiency of conversion to fuel carbon was about 43%. The reason for the uncertainty is that, although the digester had not been fed for a period of 60 days prior to the onset of Macrocystis feeding, the accumulated nutrients in the digester were being converted to gas at a significant rate at the time the Macrocystis test was started. Calculations were based on an assumption that there was a daily gas production of 6 liters (STP) due to the previously accumulated volatile solids.

This volume was deducted from the total gas produced each day. The total methane production assumed to be produced from the Macrocystis was 49.5 liters (STP), equivalent to 26.5 g carbon. The total carbon added as Macrocystis during the test period was 61 g, and therefore, an approximate conversion efficiency of 43% was obtained. The performance of each digester during the period of feeding with Macrocystis is shown in Tables 7.9 and 7.10. These tables include total gas production, and gas composition as the ratio of CH_4 and CO_2.

TABLE 7.9: PERFORMANCE OF DIGESTER 8B WITH *Macrocystis pyrifera*

Date	Total Gas Produced (l, STP)	Gas Composition CH_4/CO_2	CH_4 Produced (l, STP)
12/12/73	7.9	1.70	5.0
12/13/73	9.6	1.90	6.3
12/14/73	10.4	1.68	6.5
12/17/73	29.4	1.55	17.9
12/18/73	10.9	1.31	6.2
12/19/73	10.8	1.19	5.9
12/20/73	10.8	1.10	5.6
12/21/73	11.1	1.08	5.8
12/23/73	39.8	1.16	21.4
12/25/73	15.0	1.27	8.4
12/27/73	27.5	1.20	15.0
12/31/73	13.2	1.31	7.5

Source: PB 238 103

TABLE 7.10: PERFORMANCE OF DIGESTER 1B WITH *Macrocystis pyrifera*

Date	Total Gas Produced (l, STP)	Gas Composition CH_4/CO_2	Total Methane Produced (l, STP)
12/12/73	0.9	0.70	0.4
12/13/73	1.4	0.63	0.5
12/14/73	2.0	0.53	0.7
12/17/73	6.0	0.54	2.1
12/18/73	1.4	0.51	0.5
12/19/73	2.0	0.47	0.6
12/20/73	2.2	0.51	0.7
12/21/73	2.9	0.50	1.0
12/23/73	9.1	0.53	3.1
12/25/73	4.1	0.52	1.4
12/27/73	4.2	0.33	1.0
12/29/73	5.4	0.25	1.1
12/31/73	4.2	0.27	0.9

Source: PB 238 103

Conversion of Fresh Water Algae to Methane: Digester 6B, operating at 48°C in fresh water, was fed a mixture of algae and dog food for a period of eighteen days and then the feed was changed to 100% algae at a loading of 4.0 g total solids/liter/day and a retention time of 10 days. This regime was continued until the test was terminated. The algae used were obtained from the University of California Sanitary Engineering Research Laboratory and had been grown in a sewage oxidation pond.

The data for the test period is presented in Table 7.11. The volatile acids are moderately high and the solids content of the digester liquid is quite high. However, operation of the digester, based on gas production, gas composition and pH of the system, appears to be without problems. The carbon fed during the test period was 628 g. The amount of carbon appearing as methane during this period was 137 g. This yields a conversion efficiency of 21.8%. Although this value is somewhat low, it may be expected to rise with continued operation. The build-up of total and volatile solids in the digester could be prevented by decreasing digester loading which in turn, might increase conversion efficiency.

Conversion of Animal Wastes to Methane: Operation of one digester (No. 2A) utilizing animal waste in the form of cattle manure was continued during this period of the program. The waste material has been used either as received or after caustic treatment to increase the availability of refractory organic components.

Operation at 48°C with a loading of 4 g total solids/liter/day and with a 10 day residence time using the cattle manure as received, resulted in a conversion efficiency of animal waste to methane (based on carbon content) of 8.6%. Increasing the residence time from 10 to 20 days with a loading of 2 g total solids/liter/day gave a conversion efficiency of 25.3% based on the carbon contents of the feed and methane.

TABLE 7.11: DIGESTER 6B–CONVERSION OF FRESH WATER ALGAE TO METHANE

Date	Gas Production Rate (l, STP/day)	Gas Production Composition CH$_4$/CO$_2$	pH	Alkalinity (mg CaCO$_3$/l)	Volatile Acids (mg/l)	Total Solids	Volatile Solids	Carbon	Total Nitrogen	NH$_3$–N (mg/l)
10/8/73	36.0	1.62	7.20	4,900	1,416					
10/9/73	21.9	1.55	7.30							
10/10/73	36.1	1.36	7.48	6,310	1,630					
10/11/73	46.3	–	7.25	5,800	1,870	17,960	16,000	8,060	3,050	940
10/12/73	39.8	1.37	7.18	6,310	2,020					
10/13/73	76.8	1.42								
10/14/73			7.40							
10/15/73	30.8	1.39	7.32	6,370						
10/16/73	31.2	1.41	7.30	6,690						
10/17/73	21.9	1.63	7.32	6,310	1,700					
10/18/73	22.9	1.39	7.28	5,860		19,360	17,060	9,240	2,400	1,250
10/19/73	28.2	1.06	7.15	6,720						
10/20/73	29.6	1.44	7.17							
10/21/73	40.0	1.66								
10/22/73			7.35	6,040						
10/23/73	12.6	1.45	–							

NOTE: Loading, 4.0 grams total solids/liter/day; detention time, 10 days; temperature, 48°C.

Source: PB 238 103

Conversion of Grass to Methane: Digester 1A (6) had been operating with grass cuttings (both fresh and dried) as the only feed for about a year. During this period of operation the digester loading was varied from 1.9 to 4.5 grams total solids/liter of digester volume/day at a 10 day detention time, and from 2.75 to 4.0 grams/liter/day at 20 day residence. Data were also collected at 10 and 28 day detention time using a grass loading of 4.5 grams total solids/liter/day.

In all cases operation was at 48°C. Conversion to methane was generally good and the composition of the digester liquid was excellent. Problems related to the use of grass were due to its tendency to form a heavy floating mat which did not allow free passage of product gas out of the digester.

Digesters 4B and 2B were also operated with grass as a significant part of the feed. The feed to Digester 4B consisted of 50% grass and 50% domestic solid waste. Loading was at 4.0 g total solids/liter/day and the retention time was 10 days. Temperature of operation was 48°C. Conversion was acceptable and digester operation was smooth except for the formation of a floating mat of partially converted feed in the digester.

Digester 2B was operated with 30% grass and 70% domestic solid waste. Operation was satisfactory with respect to conversion of feed to methane, although there was a tendency for the digester to become somewhat acidic.

Digester 3B was also operated with a feed of grass and domestic solid waste, but in this case a marine environment existed within the digester. Until deactivation, a mixture of 50% grass and 50% domestic solid waste (both based on dry solids) was added to the digester at a rate of 2.0 or 4.0 g total solids/liter/day and at a detention time of 20 days. The digester was operated at 48°C. Gas production was relatively poor and the gas tended to be high in carbon dioxide. Since it has been found that long detention times are most effective with marine digesters it is expected that operation of this digester would have improved if concentrated feed slurries had been used.

Conversion of Domestic Solid Waste to Methane: Digester 3A was operated with domestic solid waste as the feed. Loadings ranging from 2.6 to 4.0 g total solids/liter/day were used with detention times between 10 and 15 days. The operating temperature was 48°C. Gas production was moderate and the gas produced was well over 50% methane. The relatively low gas production was reflected in the tendency towards a build-up of solids in the digester. It is probable that only the more easily attacked materials in the solid waste were being converted and that there was very little conversion of the paper component of the waste. A lack of adequate cellulase activity has been noted on several occasions in these laboratory digesters.

Summary of Overall Digester Operating Results: In order to provide an overview of the results obtained in converting organic materials to methane through the use of an anaerobic digestion process, the efficiency of conversion and the operating conditions are presented in Table 7.12 for the individual materials studied in this program.

The efficiency of the conversion of the organic material to methane is expressed as the amount of carbon in the form of CH_4 relative to the amount of carbon in the materials added to the digesters.

TABLE 7.12: PERFORMANCE SUMMARY OF DIGESTERS

	Digester Number	Digester Loading (g/l/day)	Detention Time (days)	Temperature (°C)	Digester Efficiency (CH$_4$-C/feed C)
(a) Fresh water systems					
Domestic solid waste (66% paper, 17% garbage, 17% grass)		4.0	10	48	19
		2.6	15	48	14
		4.0	10	48	9
Animal wastes		2.0	20	48	17
Animal waste-caustic treated		2.0	20	48	25
Fresh water algae		4.0	10	48	22
Grass		1.8	28	48	17
Water hyacinth		4.0	10	48	very low
Water hyacinth-caustic treated		4.0	10	48	6
Seaweed		2.0	20	39	very low
		2.0	20	33	very low
Seaweed-caustic treated		2.0	20	39	24
		2.0	20	33	9
Giant kelp		1.8	20	39	41
		1.8	20	33	24
(b) Seawater systems					
Domestic solid wastes		4.0	10	48	12
Giant kelp		1.0	50	48	43
(c) Fresh water					
	1	4.0	10	48	40
	3	4.0	10	48	43
	4	4.0	10	48	40
	2B	4.0	10	48	42
	4B	4.0	10	48	42
	7B	4.0	10	48	39
Seawater Converted from fresh water systems	5	4.0	10	48	24
	5	4.0	~100	48	40
	4	4.0	30	48	55
	8B	4.0	10	48	6
	1B	2.8	14	48	13
Started with effluent from Digesters 4 or 5	2	4.0	10	48	12
	2	2.0	20	48	25
	3B	4.0	10	48	11
	3B	1.4	28	48	12
	3B	2.0	20	48	9
	3B	2.8	14	48	8

Source: PB 238 103

The materials examined include domestic solid waste, animal waste, grass, water hyacinth, fresh-water unicellular algae, seaweed and giant kelp. A number of these materials were examined in a conventional fresh water environment and in a seawater environment.

From the data presented in part (a) of the table, it is seen that giant kelp, *Macrocystis pyrifera,* has been converted to CH_4 with the high efficiency of about 40%. All of the other materials studied with the exception of untreated water hyacinth and seaweed were converted to CH_4 carbon with an efficiency in the neighborhood of 20%. The more resistant hyacinth and seaweed exhibited improved conversion efficiencies after preliminary attack by dilute sodium hydroxide solution. Treatment of the cattle manure in this way also appeared to sensitize some of the more refractory components and make them susceptible to later biological attack.

The highly efficient conversion of the carbon in the giant kelp, *M. pyrifera,* was carried out in a digester containing organisms from cattle rumen in a digester started with organisms from a municipal sewage sludge digester. The efficiency of conversion was reduced to about 24%. The same digesters were used for the study of caustic-treated seaweed and similar differences in conversion efficiency were observed.

In part (b) of the table, the conversion efficiencies with domestic solid waste and with giant kelp in seawater digesters are presented. It is seen that the results are generally comparable with data obtained for the same materials in fresh water systems. Again, the very highly efficient, 43%, conversion of kelp to CH_4 is emphasized.

In part (c), comparisons are presented of the conversion efficiencies realized with six well-adapted fresh water digesters fed with a standardized organic material. Operating conditions in the six digesters were the same although the physical size of the first three digesters listed were only about one third the size of the last three. Conversion efficiencies were uniformly high and ranged from 39 to 43%.

Seawater digesters operated with the same standard feed also yielded conversion efficiencies of approximately the same magnitude. One of the digesters consistently showed an efficiency of 55% for the conversion of feed carbon to methane carbon. These digesters had been converted to a seawater environment by slow adaptation of fresh water systems. The data presented in the last part (c) shows the relatively poorer performance of digesters started with effluent from the well-operating units but in which the volume was increased to the desired level by fairly rapid dilution with seawater. These seem never to achieve the high activity of the original digesters.

The results obtained in this study have demonstrated differences among the plant materials used as feed with respect to both ease and extent of conversion to CH_4. The ability to convert about 40% of the carbon content of the giant kelp, *M. pyrifera,* to methane strongly suggests that further study should be carried out. Seawater digesters have been satisfactory in this application, and it is suggested that their use is indicated when seawater wetted marine plants are considered as a material for conversion to CH_4 by an anaerobic fermentation process.

Results of Single-Stage Digestion—University of Pennsylvania

There were four digesters operating in the single stage mode. This mode of operation is defined as the coexistence of the acidogenic and methanogenic bacteria within the confines of a single tank. Two single stage digesters operate at 37° and one at 48°C. The operating temperature of one of the thermophilic digesters, Digester 4, was 55°C to investigate higher temperatures. The method of operation of these digesters is summarized in Table 7.13. Operating data are summarized in Table 7.14.

TABLE 7.13: METHOD OF OPERATION OF SINGLE STAGE DIGESTERS

Digester#	Temperature	Dentention time
1	37°C	10 days
2-1	48°	9
4	55°	10
5	37°	10

Source: PB 238 103

TABLE 7.14: CONVENTIONAL SINGLE STAGE DIGESTION—OPERATING DATA

pH	7.0
Feed	5% dog food slurry
Loading	0.18-0.41 lb VS/ft^3/day (0.1-0.20 lb VS/ft^3/day-sewage sludge digestion)
Theoretical Dentention Time	9, 10 days
Volatile Acids Concentration	50-1300 mg/liter as acetic acid
Gas Production	7-9.0 ft^3 gas/lb VS
Gas Composition	$\frac{CH_4}{CO_2} = 1.4-1.7$
CH$_4$ Production	4.4-5.4 ft^3 CH$_4$/lb VS
Percent CO$_2$	37-42
Alkalinity	2000-3600 ppm as Ca CO$_3$

Source: PB 238 103

Digester Reaction to Batch Feeding: Important digester operational performance criteria (gas production, pH and temperature) are continuously recorded in order to evaluate digester responses to experimental conditions. A characteristic pattern of pH and gas production is observed for the single stage digesters following once daily batch feeding. This pattern is shown in Figure 7.4.

FIGURE 7.4: VOLATILE ACIDS, pH, AND GAS PRODUCTION CHANGES FOLLOWING ONCE DAILY BATCH FEED OF THERMOPHILIC SINGLE STAGE DIGESTER

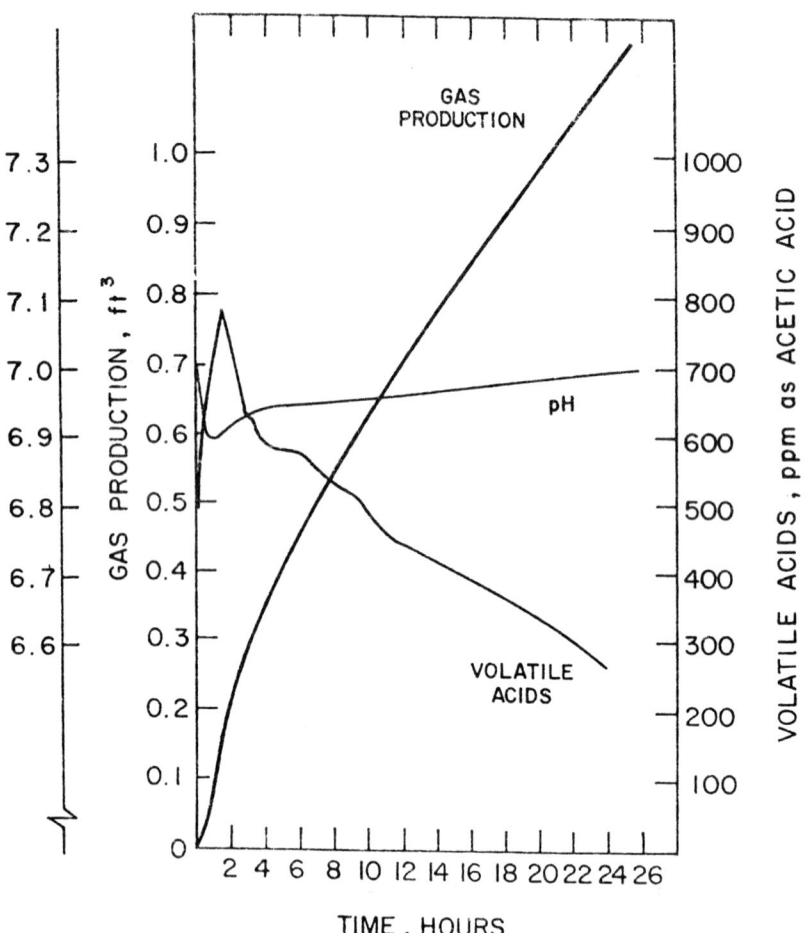

Source: PB 238 103

The data were obtained as a tracing from the continuous record of digester 2-1. The data reveal a dip in pH during the first hour, and a decreasing gas production rate following the first hour peak. This is interpreted as the stimulation of acidogenic bacteria with the resultant production of volatile acids and decrease in pH.

The implication of this observation is that the pH drop presents a stress to the methanogens. Consequently, methane production may be somewhat inhibited by the daily increase in hydrogen ion concentration. This agrees well with the rule of thumb in anaerobic sewage digestion, that it is better to feed a digester in small amounts at frequent intervals than in large amounts infrequently. That is, digester performance is improved by minimizing the (H_3O^+) fluctuations.

In order to more thoroughly evaluate the changes which occur following batch feed, an experimental program was initiated. Changes in pH, volatile acids, gas composition and gas production rates were determined before and after feeding the digesters in a once daily, twice daily and continuous fashion.

The results are presented in Figures 7.4, 7.5 and 7.6. The data were obtained with digester 2-1. The values on the ordinate corresponding to t = 0, represent the digester condition immediately prior to feeding. It is apparent that significant changes occur during the first one to two hours subsequent to batch feeding.

FIGURE 7.5: GAS PRODUCTION RATE AND $CH_4:CO_2$ CHANGES FOLLOWING ONCE DAILY BATCH FEED OF THERMOPHILIC SINGLE STAGE DIGESTER

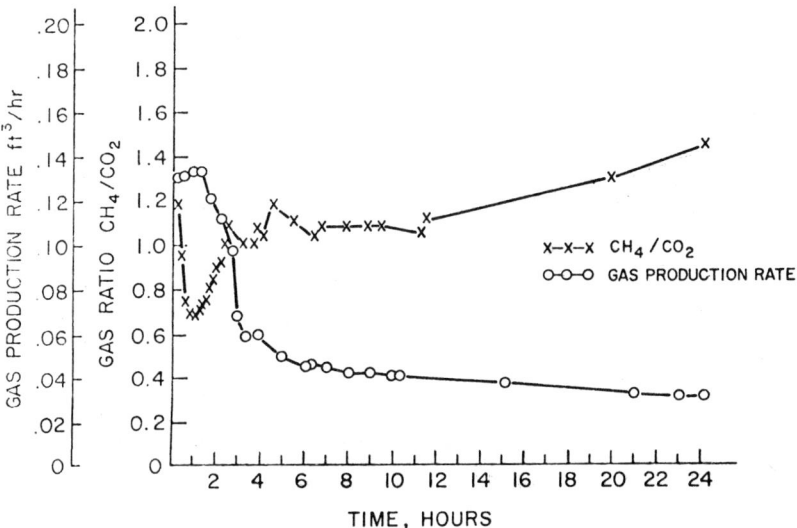

Source: PB 238 103

FIGURE 7.6: RATES OF CO_2 AND CH_4 PRODUCTION FOLLOWING DAILY BATCH FEED

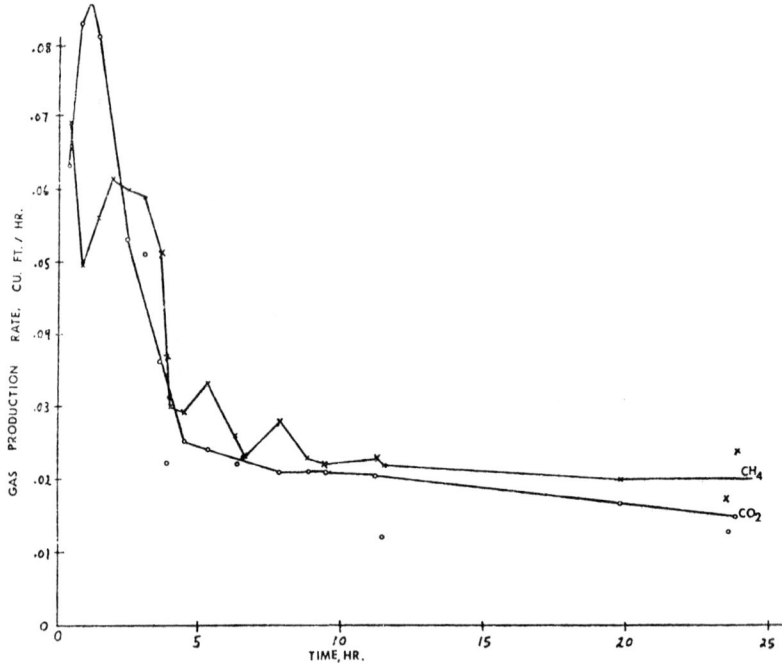

Source: PB 238 103

From Figure 7.4, it is seen that the total volatile acids (as mg CH_3COOH/l) concentration increases by approximately 60% during this time interval. A drop of about 0.1 pH units was observed on this date.

In Figure 7.5, the gas analysis results are presented. The values for gas production rate were obtained by determining the derivative of the gas production curve (Figure 7.4) at the points for which the gas analyses were made. The partial pressure of CH_4 immediately drops coincident with an increase in pCO_2. The ratio of pCH_4/pCO_2 reaches a minimum at t = 1 hr. Total rate of gas production (ft^3/hr) remains relatively high, approximately 0.13 ft^3/hr for two hours, and then decreases dramatically to 0.06 ft^3/hr followed by a gradual decrease to 0.04 ft^3/hr.

Production rates of CO_2 and CH_4, following the once daily batch feed are presented in Figure 7.6. The rates were calculated utilizing the expression,

$$C = C_0 + \frac{C_1 - C_0}{1 - e^{-Rt/V}}$$

where C is the partial pressure of the gas being produced; C_0 and C_1, the measured partial pressures at the beginning and end of the time period t; R is the total gas production rate (ft^3/hr); and V, the dead volume (ft^3) of the gas space in the digester. This equation was derived to account for the dilution of the produced gas within the volume of gas previously produced. The rate of production of a particular gas was then calculated as CR.

The rate of CO_2 production increased immediately after feeding and reached a peak at t = 2 hours (Figure 7.6). Thereafter, the CO_2 production rate dropped rapidly to 0.25 ft^3 CO_2/ hr. The pattern exhibited by the CO_2 generation rate is interpreted to indicate the rapid onset of the metabolic activities of the acid producing bacteria.

The CH_4 production rate (Figure 7.6) follows a similar pattern of increased production rate and then a rapid decrease to 0.3 ft^3/hr. An additional factor is the dip in the CH_4 production rate curve at t = 1 hour. One explanation for this is that the earlier peak, at t = 0.5 hr, is due to the production of CH_4 directly from CO_2 and H_2. Although H_2 is rarely observed in the single stage digesters, it is produced in the first stage of the two stage digester (pH_2 = 0.02 to 0.1). Thus, it may be that there is an early burst of hydrogen production which is rapidly converted to CH_4. Following the consumption of the H_2, methane production decreases sharply until other methane producing mechanisms begin to operate.

The results obtained confirmed earlier interpretations that the changes observed immediately after feeding are related to the rapid increase in the metabolism of the acidogenic bacteria. If the hypothesis that these changes place stress on the methanogenic bacteria is correct, then gas productivity (ft^3 gas/lb VS added) ought to increase as the feeding schedule approaches continuous feed. Hence, one would expect the methane productivity to increase with a continuous feeding program.

Comparison of Thermophilic and Mesophilic Digestion: The role of temperature in the regulation of digester efficiency is known to be important. Data has been collected at 37°C (mesophilic) and 48° and 55°C (thermophilic). As reported earlier, the mesophilic digesters produce more gas (ft^3 gas/lb dry organic matter added) than the thermophilic digesters. The average value for the mesophilic digesters is 8.59 ft^3/lb dry organic matter added/day, and, for the thermophilic digesters, 7.92 ft^3/lb VS/day. These values represent average gas productivities obtained during periods of constant feed.

An experimental program has been undertaken to evaluate the effect of temperature and loading rate on gas productivity. One major purpose is to determine the maximum loading at various temperatures and the resultant gas production. The data collected are presented in Tables 7.15 and 7.16 and summarized in Tables 7.17 and 7.18.

The statistical information presented in these tables includes the arithmetic standard deviation, the number of data points included in the mean, and the coefficient of variation (standard deviation divided by the mean), CV, expressed as a percentage.

From these data, it can be seen that the general relation between gas production at 48°C and 37°C holds for various loadings. That is, in general, the production of gas, in terms of ft^3 of gas produced/lb dry organic matter added, is greater at 37°C than 48°C. This agrees with data presented by Pfeffer (5) indicating that productivity is greater at 35°C than at 45°C. Thus, regarding methane

TABLE 7.15: CARBON CONVERSION EFFICIENCY FOR SINGLE STAGE MESOPHILIC DIGESTERS

Loading (lb VS/ft³/day)	Gas Production (ft³ gas/lb VS/day)	Carbon in/day (grams)	Carbon out in gas phase/day (grams)	Carbon Conversion (percent)
0.181	8.50	13.5	9.1	67.4
0.188	8.54	14.0	9.4	67.1
0.211	8.70	15.7	10.8	68.8
0.235	8.33	17.5	11.5	65.7
0.258	8.18	19.2	12.4	64.6
0.286	8.63	21.3	14.5	68.1
0.290	8.88	21.6	15.1	69.9
0.311	8.98	23.1	16.5	71.4
0.352	8.40	26.2	17.4	66.4

TABLE 7.16: CARBON CONVERSION EFFICIENCIES FOR SINGLE STAGE THERMOPHILIC DIGESTERS

Loading (lb VS/ft³/day)	Gas Production (ft³ gas/lb VS/day)	Carbon in/day (grams)	Carbon out in gas phase/day (grams)	Carbon Conversion (percent)
0.240	8.46	19.5	13.1	67.2
0.247	8.15	20.1	12.9	64.2
0.258	7.90	20.9	13.2	63.2
0.260	7.67	21.1	12.9	61.1
0.284	7.25	23.1	13.2	57.2
0.288	7.37	23.4	13.7	58.6
0.312	8.20	25.3	16.4	64.8
0.344	7.67	27.9	17.0	60.9
0.365	8.12	29.7	19.1	64.4
0.408	8.12	33.2	21.4	64.5

TABLE 7.17: ENERGY CONVERSION EFFICIENCIES FOR SINGLE STAGE MESOPHILIC DIGESTERS

Loading (lb VS/ft³/day)	Gas Productivity (ft³ gas/lb VS/day)	Input Energy (Btu/lb VS)	% CH₄	Output Energy (Btu/lb VS)	Energy Conversion Efficiencies
0.181	8.50	9,450	60	5,100	54.0
0.188	8.54			5,120	54.3
0.211	8.70			5,220	55.3
0.235	8.33			5,000	52.9
0.258	8.18			4,910	52.0
0.286	8.63			5,170	54.8
0.290	8.88			5,330	56.8
0.311	8.98			5,390	57.0
0.352	8.40			5,040	53.3

Source: PB 241 055

TABLE 7.18: ENERGY CONVERSION EFFICIENCIES FOR SINGLE STAGE THERMOPHILIC DIGESTERS

Loading (lb VS/ft³/day)	Gas Productivity (ft³ gas/lb VS/day)	Input Energy (Btu/lb VS)	% CH$_4$	Output Energy (Btu/lb VS)	Energy Conversion Efficiencies
0.240	8.46	9,450	60	5,070	53.7
0.247	8.15			4,890	51.7
0.258	7.90			4,740	50.1
0.260	7.67			4,600	48.7
0.284	7.25			4,350	46.1
0.288	7.37			4,420	46.8
0.312	8.20			4,920	52.1
0.344	7.67			4,600	48.7
0.365	8.12			4,860	51.5
0.408	8.12			4,860	51.5

Source: PB 241 055

production, a more economical approach is mesophilic digestion. Under this condition, operating costs are reduced due to the maintenance of a lower temperature, and there is a potential for greater methane productivity.

The thermophilic digester operating at 55°C was not brought to full loading (5 g VS/liter/day). The data collected at lower loadings indicate that the gas productivity is 8.54 scf gas/lb VS added/day (N = 47, S = 4.51). This value is greater than that observed with the 48°C thermophilic digesters, and is in the range of results for the 37°C mesophilic digesters. These data, obtained with particulate simulated solid waste, agree with the conclusions of Pfeffer (5).

The volatile solids destruction data indicate a relationship with temperature, detention time and loading rate. For the particulate feed used, the destruction of volatile solids increased with decreasing loadings, increased with temperature: 55 ≈ 37 > 48°C, and with increasing detention times. The data are summarized in Table 7.19.

TABLE 7.19: SUMMARY OF VOLATILE SOLIDS DESTRUCTION DATA

Loading	Detention time	Thermophilic 48°C	Thermophilic 55°	Mesophilic 37°
.188	10			71.5
.312	10	63.3		
	9	61.4		
.352	10			56.7
.408	10	58.9		
Variable:				
-Periods of constant feed	10	60.9		59.3
-all data	10	65.8	81.1	67.6

Source: PB 238 103

The mesophilic digesters in these experiments have been subject to more instability problems than have the thermophilic cultures. Greater difficulties have been encountered in pH maintenance with the mesophilic systems than with the thermophilic. This has necessitated the addition of lime and reduction in feed rates.

Conversion Efficiency: The efficiency of anaerobic digestion may be expressed in terms of the percent of the input carbon which appears in the gas phase. Carbon conversion efficiencies for the single stage digesters are presented in Tables 7.20 and 7.21. The calculations were made using the value of 38.1% carbon in the feed material, as determined experimentally at the United Aircraft Research Laboratories. This value was utilized to calculate the input carbon. The output carbon was calculated by converting the gas production rate (from Tables 7.17 and 7.18) to grams of carbon.

TABLE 7.20: CARBON CONVERSION EFFICIENCY FOR SINGLE STAGE MESOPHILIC DIGESTERS

Loading (lb VS/ft^3/day)	Gas Productivity (ft^{3}gas/lb VS/day)	g C in/day	g C out in gas phase/day	% Carbon conversion
0.181	8.50	13.5	9.1	67.4
.188	7.90	14.0	8.8	62.9
.211	8.70	15.7	10.8	68.8
.235	8.33	17.5	11.5	65.7
.258	8.18	19.2	12.4	64.6
.286	8.63	21.3	14.5	68.1
.290	8.88	21.6	15.1	69.9
.311	8.98	23.1	16.5	71.4
.352	8.68	26.2	18.0	68.7

TABLE 7.21: CARBON CONVERSION EFFICIENCIES FOR SINGLE STAGE THERMOPHILIC DIGESTERS

Loading (lb VS/ft^3/day)	Gas Productivity (ft^{3}gas/lb VS/day)	g C in/day	g C out in gas phase/day	% carbon conversion
.240	8.46	19.5	13.1	67.2
.247	8.15	20.1	12.9	64.2
.258	7.90	20.9	13.2	63.2
.260	7.67	21.1	12.9	61.1
.284	7.25	23.1	13.2	57.2
.288	7.37	23.4	13.7	58.6
.312	8.43	25.3	16.9	66.8
.344	7.67	27.9	17.0	60.9
.365	8.12	29.7	19.1	64.4
.408	7.40	33.2	19.4	58.4

Source: PB 238 103

An alternate procedure for estimating efficiency is in terms of energy conversion. These data are presented in Tables 7.22 and 7.23. The input energy calculation is based on a feed material energy content of 5.25 kcal/g, or

TABLE 7.22: ENERGY CONVERSION EFFICIENCIES FOR SINGLE STAGE MESOPHILIC DIGESTERS

Loading (lb VS/ft³/day)	Gas Productivity (ft³ gas/lb VS/day)	Input Energy (Btu/lb VS)	% CH_4	Output Energy (Btu/lb VS)	Energy Conversion Efficiencies
0.181	8.50	9,450	60	5,100	54.0
0.188	7.90			4,740	50.2
0.211	8.70			5,220	55.3
0.235	8.33			5,000	52.9
0.258	8.18			4,910	52.0
0.286	8.63			5,170	54.8
0.290	8.88			5,330	56.8
0.311	8.98			5,390	57.0
0.352	8.68			5,210	55.1

TABLE 7.23: ENERGY CONVERSION EFFICIENCIES FOR SINGLE STAGE THERMOPHILIC DIGESTERS

Loading (lb VS/ft³/day)	Gas Productivity (ft³ gas/lb VS/day)	Input Energy (Btu/lb VS)	% CH_4	Output Energy (Btu/lb VS)	Energy Conversion Efficiencies
0.240	8.46	9,450	60	5,070	53.7
0.247	8.15			4,890	51.7
0.258	7.90			4,740	50.1
0.260	7.67			4,600	48.7
0.284	7.25			4,350	46.1
0.288	7.37			4,420	46.8
0.312	8.43			5,060	53.5
0.344	7.67			4,600	48.7
0.365	8.12			4,860	51.5
0.408	7.40			4,440	47.0

Source: PB 238 103

9,450 Btu/lb VS. This value was calculated based on a feed material composition of 40.5% protein at 5.7 kcal/g dry VS, 8% fat at 9.5 kcal/g and 51.5% carbohydrate at 4.2 kcal/g dry VS.

Volatile solids reduction is also utilized as a measurement of conversion efficiency. The volatile solids content of the feed is 90.3% (dry basis). The average destruction of volatile solids in the 37°C mesophilic digesters was 67.6% (N = 47, S = 19.2), and, for the thermophilic digesters, 65.8% (N = 50, S = 15.1). (See Table 7.19.)

The data reflect the increased efficiency of mesophilic digestion over themophilic digestion. The lack of a clear-cut inverse relationship between loading and efficiency for the digesters implies that increased loadings may be applied without sacrifice of conversion efficiency. The volatile solids data, however, are not in agreement with this implication.

At the current loadings, approximately 0.4 lb VS added/ft³ culture volume/day, the volatile acids concentrations are 250 and 1,450 mg CH_3COOH/l, for the mesophilic and thermophilic systems, respectively. At the same time, the alkalinities, expressed as $CaCO_3$ are 3,265 and 3,325 mg/l, respectively. Expressing the volatile acids values as $CaCO_3$, one obtains 420 and 2,410 mg/l. In each case, the alkalinity exceeds the volatile acids, demonstrating stability as recommended by Dague (6). This is an additional indication that the loadings may be successfully increased. This means that the size of the digesters, for a given energy production rate, may be reduced. Since the capital cost of the digester represents a significant factor in the total cost picture, the economic implication is for reduced methane production costs.

Two Stage Digestion

In this context, two stage digestion refers to the separation of the acidogenic bacteria and the methanogenic bacteria into two separate tanks operated in series. The first stage is characterized by the accumulation of acids, and, it is in the second stage that the majority of the methane is produced.

Stage 1: The first stage was operated successfully for many months. The digester was initiated by overloading a conventional single stage digester and by holding the pH at 6 by the automatic addition of lime. pH control presently requires 1 liter/day of a lime slurry containing 35.2 g $Ca(OH)_2/l$. The current operational data for the Stage 1 digester are summarized in Table 7.24.

TABLE 7.24: STAGE 1 DIGESTER—OPERATING DATA

pH	6.0, maintained with $Ca(OH)_2$
Feed	100 g dry organic matter/day (as 5% slurry)
Loading	0.52 lb VS/ft^3/day
Theoretical detention time	6 days
Volatile acids	13,000 ppm as acetic acid
Gas production	1.9 ft^3 gas/lb dry organic matter
Gas composition	98% CO_2
Alkalinity	12,000 ppm as $CaCO_3$

Source: PB 238 103

The Stage 1 digester has been operated in various modes since its inception. At first, Stage 1 effluent, without solids separation, was transferred to the second stage. During this period, the detention time was five days. Later, solids were not transferred to the second stage. Detention time during this phase was at first five days, and then it was reduced to four days. The solids separation step occurred within the Stage 1 tank, by allowing the solids to settle to the seven liter mark (total culture volume is 12 liters), followed by transfer of 2 liters of supernatant and by the drainage of 1 liter of settled solids from the bottom of the tank.

The lack of solids transfer was associated with decreased second stage gas production. Consequently, solids transfer was resumed. This phase was conducted at detention times of four and six days. Gas production data during the different operational modes are presented in Table 7.25. Other data are summarized in Table 7.26. Gas production has increased during the time period of these changes. The interval during which the loading was variable represents the adaptation of the culture to Stage 1 operation.

TABLE 7.25: STAGE 1 GAS PRODUCTION

Loading lbVS/ft^3/day	Detention Time	Solids Separation Within Tank		Gas Production, ft^{3}gas/lb VS added		
		Yes	No	mean	N	std.dev.
Variable	5 days		no	2.39	16	1.41
0.473	5 days		no	1.10	32	0.43
0.473	5 days	yes		1.35	11	0.95
0.519	4 days	yes		3.26	24	4.50
0.519	4 days		no	1.21	42	.54
0.519	6 days		no	1.92	94	.95
0.519 (22°C)	6 days		no	1.73	10	2.59

TABLE 7.26: TWO-STAGE OPERATING CHARACTERISTICS*

Stage	Temperature	Detention Time	Volatile Acids mg/l as acetic	(N)	Alkalinity mg/l as CaCO$_3$	(N)	% VS destruction	(N)
1	48	4	7,843	(7)				
2	48	10	128	(7)				
1	48	6	13,411	(20)	12,463	(12)	82	(22)
2	48	10	277	(20)	4,391	(11)	95	(22)
1	22	6	13,015	(2)			83.6	(2)
2	48	10	192	(2)			94	(2)

*All data are for solids transfer to Stage 2 and a Stage 1 loading of 0.519 lb VS added/ft^3 culture volume/day.

Source: PB 238 103

During the time span indicated in Table 7.25, the composition of the produced gas changed from about 98% CO_2 and 2% H_2 at the end of the phase without solids separation to about 90% CO_2 and 10% H_2 during the solids separation phase. During the return to the solids transfer regime, the hydrogen production decreased, again to a value yielding a partial pressure 0.02. The production of H_2 in the first stage represents a potential source of CH_4 which could be produced by the direct metabolism of CO_2 and H_2. The bacteria which undergo this metabolic pathway are obviously inhibited in the Stage 1 tank, most likely because of the low pH.

The H_2 escaping in the exit gases is an energy loss which might be overcome. Minimization of the H_2 loss may be achieved using two approaches. The first approach is to recycle the Stage 1 gases to the Stage 2 tank where the hydrogen will become available to the methanogens present in the second stage. Gas recycle will also act as a mixing mechanism within Stage 2.

The second approach is to start up another Stage 1 tank in which a different microflora is developed. This second type of Stage 1 culture is characterized by a short detention time, pH 7, the accumulation of volatile acids, and the production of CH_4 with no H_2 in the exit gases. In order to test the applicability of this type of culture to the two-stage system, new Stage 1 cultures should be developed by reducing the detention time of single stage digesters to 3 to 5 days. The advantages of this type of Stage 1 digester would be (1) the increase in CH_4 production due to conversion of H_2; and (2) the decrease in operating costs because there would be no need to add lime.

The first stage digester has operated with considerable ease of operation since its initiation. The only problem occurred when the temperature control mechanisms malfunctioned, and the culture temperature rose from 48° to 74°C overnight. The culture was reestablished at full loadings, 0.519 lb VS added/ft^3 culture volume/day, within one week.

Operation of the first stage digester has been investigated at reduced temperatures. The digester has been run at approximately 22°C for ten days. During this time period, the gas productivity decreased slightly to 1.73 scf/lb VS added/day, whereas the volatile acids remained constant, averaging 13,015 ppm as acetic acid (Table 7.26). This indicates that the first stage can be operated at reduced temperatures without sacrifice of volatile acids productivity.

Stage 2: The second stage was operated also for many months. A conventional single stage digester was converted to the Stage 2 mode by transferring effluent from the first stage to the second stage, in lieu of the regular particulate feed which was then added only to the first stage. The initial effort to start up the Stage 2 digester failed, because excessive amounts of Stage 1 effluent were transferred. The second attempt was initiated by the transfer of 200 ml of Stage 1 effluent. The volume of effluent transferred was gradually increased to 2 liters/day. This process took about 3 weeks.

The Stage 2 feed has contained 8,000 to 14,000 mg volatile acids as acetic acid/liter, i.e., 16 to 28 g volatile acids/day. The volatile acids in the feed which have been identified include acetic, propionic, butyric, isobutyric, valeric and isovaleric. Occasionally, formic acid is observed. In addition, a peak beyond valeric acid is noted. This peak most probably represents caproic acid. The predominant volatile acid is propionic acid.

The sum of the concentrations, expressed as acetic acid, of the individual volatile acids as determined by gas chromatography is less than the concentration of total volatile acids as determined by the silicic acid, chloroform-butanol extraction chromatographic procedure. This is undoubtedly due to the presence of other acids such as lactic and succinic. These acids are included in the silicic acid method but have not been measured using gas chromatography. The presence of these nonvolatile acids is anticipated in the Stage 1 digester on the basis of the known acid producing fermentation patterns. It should also be noted that the probability that ethanol is present in the first stage is quite high.

The Stage 2 digester has been operated in several ways, the exact manner depending on Stage 1 operation. Table 7.27 summarizes the current operational data, whereas Table 7.28, presents the gas production data for the different operating conditions. Gas production for the second stage is expressed as ft^3 gas produced/lb dry organic matter added to the first stage. The gas production values are corrected for the volume of culture transferred from Stage 1 to Stage 2. Additional operating characteristics are found in Table 7.27.

TABLE 7.27: STAGE 2 DIGESTER—OPERATING DATA

pH	7.4–7.5
Feed	26,800 mg volatile acid/day
Theoretical detention time	10 days
Volatile acids concentration	50 to 400 mg/l as acetic acid
Gas production	4-6 ft^3 gas/lb dry organic matter fed to Stage 1
Gas composition	$\frac{CH_4}{CO_2} \approx 4.0$
Percent CO_2	20
Alkalinity	3,300–4,800 ppm as $CaCO_3$

TABLE 7.28: STAGE 2—GAS PRODUCTION

Loading	Solids Transferred from Stage 1 to Stage 2	Detention Time (days) 1st Stage	Detention Time (days) 2nd Stage	Gas Production (ft^3 gas/lb VS added to Stage 1) Mean	N	Std Dev	CV*
Variable	yes	5	variable	8.44	13	1.07	12.7
Variable	no	5	variable	5.27	11	0.48	9.1
100 g VS to Stage 1/day	no	5	10	4.85	5	0.38	7.8
100	no	4	10	3.58	24	0.56	15.7
100	yes	4	10	4.69	42	0.47	10.0
100	yes	6	10	5.96	94	0.56	9.4
100	yes	6	10	5.64	10	2.59	46.0

*Culture volume.

Source: PB 238 103

It is apparent from the data presented in Table 7.28 that the transfer of solids from Stage 1 to Stage 2 enhances gas production. The composition of the gas is not dependent upon the transfer of solids. That is, in either case, pCH_4 in the produced gas is about 0.8. Thus, in terms of methane production, it seems important that the solids be transferred from the first stage to the second. The most reasonable explanation for this is that the solids provide additional surface area for attachment of the necessary bacteria.

First stage detention time is also observed to affect the efficiency of the second stage. Examination of the data presented in Table 7.28 shows that as first stage detention time increases, second stage gas productivity (ft^3 gas/lb VS added to Stage 1/day) also increases. This is because of more complete breakdown of the organic feed material. Confirmation of this is apparent in the volatile acids data. First stage volatile acids concentrations increased from about 8,000 mg/l to 10,000 mg/l to 13,400 mg/l when the first stage detention time was increased from four to five to six days.

Thus, the substrate available for methanogenesis is greater at longer first stage detention times. Second stage volatile acid pool size also increased with increasing first stage detention time: from 128 to 277 ppm as detention went from 4 to 6 days.

During the time period in which the first stage has been operated at approximately 22°C, the second stage operating characteristics remained essentially constant. Gas productivity decreased slightly from 5.96 to 5.64 scf/lb VS added/day, as did the volatile acids concentration (from 277 to 192 ppm as acetic acid). The observed destruction of volatile solids did not change. The pertinent data are summarized in Tables 7.26 and 7.28.

The increased pCH$_4$ observed in the Stage 2 tank (relative to single stage digestion) may have an explanation other than increased efficiency due to separation of the stages. The increased partial pressure of methane in the second stage gas (0.8 atm/atm) as opposed to that in the single stage gas (0.60 atm/atm) may be explicable on the basis of the large amount of alkalinity transferred from the first stage to the second stage.

The concentration of alkalinity in the second stage feed is 12,500 mg/l as CaCO$_3$. This results in the relatively high pH (7.4 to 7.5) observed in Stage 2. At this elevated pH, the carbonic acid equilibrium will be pushed to the right, and more CO$_2$ will be found in the aqueous phase.

Two-Stage Conversion Efficiency: The efficiency of two-stage operation may be assessed in terms of percent carbon and percent energy conversion. The results are summarized in Tables 7.29 and 7.30. The calculations were performed as indicated earlier in the single stage conversion efficiency discussion. The heat value of hydrogen was taken to be 345 Btu/scf H$_2$. The data used in these calculations were those obtained during the periods of constant loading: 100 g VS added to Stage 1/day and 2 liters of Stage 1 effluent transferred to Stage 2/day.

The solids conversion efficiency for the two-stage system has been determined. During the period of solids transfer, the second stage effluent contained an average of 67.7% FS. This corresponds to overall percent volatile solids destruction of 95%. It should be noted that the solids data are subject to interference due to the presence of CaCO$_3$ which may interfere with the volatile residue procedure.

TABLE 7.29: TWO STAGE CONVERSION EFFICIENCY*

Detention Time, days		Temp	Solids transferred to stage 2.	Gas productivity ft^3/lb VS added/day		% carbon conversion
Stage 1	Stage 2	Stage 1		Stage 1	Stage 2	
4	10	48	no	3.26	3.58	51.6
4	10	48	yes	1.21	4.69	46.5
6	10	48	yes	1.92	5.96	62.1
6	10	22	yes	1.73	5.64	58.0

*First stage loading, 0.519 lb VS added/ft^3 culture volume/day; 2 liters transferred to Stage 2.

TABLE 7.30: TWO-STAGE ENERGY CONVERSION EFFICIENCY*

Detention time, days		Temp.	Solids transferred to Stage 2	Gas productivity ft^3/lb VS added/day		% energy conversion
Stage 1	Stage 2	Stage 1		Stage 1-H_2	Stage 2-CH_4	
4	10	48°C	no	0.33	2.86	31.4
4	10	48	yes	.02	3.75	39.8
6	10	48	yes	.04	4.76	50.5
6	10	22	yes	.03	4.51	47.7

*First stage loading, 0.519 lb VS added/ft^3 culture volume/day; 2 liters transferred to Stage 2.

Source: PB 238 103

Comparison of Single-Stage and Two-Stage Digestion: The data collected indicate that the performance of the experimental two-stage system is approximately equivalent to that obtained with conventional single-stage digestion operated under similar conditions. The comparison between single and two-stage digestion is summarized in Table 7.31.

In Table 7.31, the data for Stage 2 regarding the amount of dry organic matter added refer to the Stage 1 feed. It should be noted that the calculations are based on the average observed gas production values for the respective digesters. The data for the thermophilic single stage digester are based on 368 points with a standard deviation of 1.18, and for the thermophilic two-stage system, on 94 points with a standard deviation of 0.56.

TABLE 7.31: COMPARISON OF SINGLE STAGE AND TWO STAGE DIGESTION

Digester	Temperature (°C)	Loading (lb VS added/ ft^3/day)	Detention Time (days)	Methane Productivity (ft^3 CH$_4$/lb VS added/day)	Carbon Conversion (percent)	Energy Conversion (percent)
Single stage	48		10	4.97	62.2	49.8
Two stage:						
Stage 1	48	0.519	6	0	14.9	0.14
Stage 2	48	0.312	10	4.77	47.2	50.5
Stages 1 + 2	48		16	4.77	62.1	50.5
Stage 1	22	0.519	6	0	13.4	0.13
Stage 2	48	0.312	10	4.51	44.6	47.6
Stages 1 + 2			16	4.51	58.0	47.7

Source: PB 238 103

One principal advantage of the two-stage system is its increased stability. The pH of the Stage 2 culture has always been quite high (approximately 7.4); the volatile acids, low (approximately 400 mg/l as acetic acid); and the alkalinity, very high (3,300 to 4,400 mg/l as CaCO$_3$). The stability of the two-stage system is reflected in these data.

The economic aspects of two-stage digestion were not considered because of incomplete data. It will be necessary to consider the cost of the additional tank and the cost of the neutralizing agent (lime). Cost advantages will result in the gas clean-up step. Operating costs for gas purification will be less for the two-stage system by virtue of the increased pCH$_4$ relative to that of the single stage digesters. Alternatively, it may be stated that the Stage 2 off-gas has a higher Btu content (800 Btu/scf) than that from the single stage digesters (600 Btu/scf). The improved stability of operation is an additional cost factor to consider in the final analysis.

A check on the overall accuracy of the data was performed by evaluating a solids balance on the digesters. The input solids was determined from the residue analysis of the feed material (100 g of raw feed material contains 82.8 g VS, 819 g FS and 8.3 g water), and, for digester 3-B, the weight of Ca(OH)$_2$ added for pH control. Output solids consist of the gaseous phase and the slurry effluent. The gases leaving the system include CH$_4$, CO$_2$, H$_2$S and H$_2$O, and, in the case of digester 3-B, H$_2$.

The weights of these products were determined from the total gas production, and the partial pressures of the respective gases. Solids withdrawn with the slurry effluent have been measured by residue analysis. The specific gravity of the slurry was estimated by assuming the specific gravities of the volatile and fixed fractions to be 1.0 and 2.5, respectively.

A summary of the materials balances is presented in Table 7.32. The results indicate that the output, i.e., recovered, materials account for 91 to 104.5% of the input. The average recovery is 97.5%. The variability and range of these

TABLE 7.32: MATERIALS BALANCE FOR LABORATORY-SCALE DIGESTERS*

	...Input Material...		.... Output Material					Slurry	..Totals..		Recovered
Digester	Feed	pH Control	CH_4	CO_2	H_2	H_2O	H_2S	Effluent	In	Out	(percent)
1	68.8	0	13.4	24.3	0	0.8	0.5	25.6	68.8	64.6	94.0
1	36.7	0	6.9	12.4	0	0.9	0.2	14.3	36.7	34.7	94.5
2-1	77.6	0	16.5	29.8	0	1.0	0.6	28.9	77.6	76.8	98.9
2-1	77.6	0	17.0	30.7	0	1.0	0.6	31.8	77.6	81.1	104.5
4	87.0	0	15.5	28.5	0	1.1	0.6	33.5	87.0	79.2	91.0
5	68.8	0	14.6	26.8	0	0.9	0.5	23.4	68.8	66.2	96.2
2-2	125.7	0	33.4	23.9	0	1.0	0.2	61.0	125.7	119.5	95.1
3-B	110.7	35.2	0	22.7	0.02	1.9	0.8	125.7	145.9	151.1	103.6
Two stage overall	110.7	35.2	33.4	46.6	0.02	2.9	1.0	61.0	145.9	144.9	99.4

*All units are grams/day unless otherwise indicated.

Source: PB 238 103

data are due primarily to random errors assosicated with the collection of the basic data: gas output volumes, slurry input volume, pH control, input slurry volume, output slurry volume, residue weight, gas partial pressures, etc.

Mariculture Investigations

A program was carried out to investigate large scale ocean farming for the production of an energy crop and the conversion of the plants to fuel. The program consisted of an analytical investigation that included a search of the literature and evaluation of the data in areas of ocean site characteristics, plant properties, hydrodynamics, fertilizer production, nutrient diffusion, ocean equipment, and mechanical processing. The data were assimilated into three growing models and an economic evaluation was made for each model.

The offshore Current Floating System is based on the use of the California Current as a growing site for plants that are immersed at the source and grow while floating. The current is seeded along an 80 mile swath and the crop spreads to a 400 mile swath while growing. The central 250 miles which contains 80% of the crop is harvested. The yield for a single passage system is 1.3×10^{15} Btu/yr and the cost of producing fuel is $4.85 per million Btu. A multiple passage system in which the plants remain in the water when the current reverses direction can double the yield and reduce the cost to $2.50 per million Btu.

The Deep Ocean Floating System offers larger growing areas, but the cost of transporting the crops for long distances was substantial. The cost of fuel was $4.80 per million Btu. The Offshore Anchored System uses the continental shelf off the east coast of the United States as a growing site. To use sandy bottoms and 300 foot depth, an artificial mooring grid was supplied for the plants. The potential yield is 7×10^{15} Btu/yr at a cost of $3.00 per million Btu.

The analytical program was divided into four studies. In site studies, physical and chemical oceanographic data were reviewed for the California Current, the Peru Current, the Sargasso Sea, the western portion of the north subtropical

Pacific Ocean, and the continental shelf off the east coast of the United States. The total ocean area covered by the study was approximately 2,000,000 square miles. The problem of competition with other commercial and recreational uses of the water was considered and some of the legal aspects of mariculture were reviewed.

In plant studies, data were gathered for the three phyla which were considered most promising as energy crops. They are Chlorophyta, Phaeophyta, and Rhodophyta. The available properties were transformed into ranking parameters such as carbon/nitrogen ratio which expresses the probability of high conversion efficiency in anaerobic digestion. The weight of methane produced per unit weight of wet plant was also derived because it affects the overall system.

Phaeophyta offer the greatest potential for ocean farming, particularly those in the order Laminariales. Laminariaceae are reported in some literature to have high productivity and some of the plants have a high solids content. The growing characteristics of some Lessoniaceae have been studied in detail, particularly Macrocystis which is harvested commercially.

The study of plant nutrition revealed data that dealt with the effects of various chemicals including nitrogen, phosphorus, boron, bromine, calcium, cobalt, copper, iodine, iron, lithium, magnesium, manganese, molybdenum, potassium, rubidium, strontium, sulfur, vanadium, zinc, vitamin B_{12}, and some hormones. However, the actual nutrient requirements of large scale ocean farming are mostly unknown.

The conclusion derived from this program is that solar energy conversion through mariculture deserves consideration as a prime source of energy. The assumptions made in this study should be verified, including the concept of large floating crops, the adaptability of the east coat continental shelf as a habitat for giant kelp, and the uptake of artificial fertilizers. Conceptual designs and preliminary evaluation should be undertaken for cheap, reliable mooring and anchors for plants, and for large scale seeding and harvesting equipment.

Conclusions

Based on examination of the biological conversion of a number of materials to methane, it can be concluded that:

1. Certain plant materials are readily subject to bacterial attack and are converted to methane in good yields. The most satisfactory materials for conversion to methane were (a) fresh water algae (Chlorella) grown in a sewage oxidation pond and (b) a marine giant kelp (*Macrocystis pyrifera*) harvested from the Pacific Ocean off the California coast.
2. *Macrocystis pyrifera* conversion was examined in both fresh water and seawater. Conversion was satisfactory in each environment. Differences might be explained as variations in the bacterial populations present in the individual digesters.
3. Other plant materials were resistant to attack by methane producing bacteria. These materials were water hyacinth and common seaweed (*Ascophyllum nodosum*).

(continued)

4. Improved conversion of the resistant materials was attained by a hydrolytic pretreatment. Either treatment with the enzyme cellulase or with hot caustic solution increased the extent to which the materials were converted to methane.
5. Animal waste conversion to methane was fairly effective with untreated waste. Pretreatment of the waste with hot caustic appeared to improve the conversion efficiency.
6. Conversion efficiencies achieved in this study have been as high as 36% when calculated on the basis of carbon in the feed and carbon in the methane produced.
7. A high ash content in *Macrocystis pyrifera* made the conversion relatively low on the basis of total solids. Performance of fresh water algae (Chlorella) was more satisfactory on this basis.
8. The biological conversion of freshly harvested plant materials to methane appears to be a viable route to production of storable fuel since the process is directly operative with plants having a very high water content.
9. Efficiency of single-stage digester operation is greater in the mesophilic ($37°C$) mode than in the thermophilic ($48°C$) mode. This has been determined in terms of gas production, methane productivity, percent volatile solids destruction, percent carbon conversion and percent energy conversion. Thermophilic digesters appear to be more stable than mesophilic digesters.
10. The results indicate that single-stage digester operation under thermophilic ($55°C$) conditions is approximately equivalent to that under mesophilic ($37°C$) conditions.
11. Methane productivity from the two-stage digestion system (4.77 ft^3 methane/lb VS added to the first stage/day) is approximately equal to that from the single-stage system under essentially equivalent operating conditions.
12. The first stage of the two-stage system is operable at lower temperatures ($22°C$) without loss of efficiency.
13. The heat content of the off-gases is improved utilizing two-stage digestion (800 Btu/scf). Heat content of the product gas from single stage digesters is 600 Btu/scf.
14. The two-stage process has proven to be stable as indicated by the lack of upsets for this system.

REFERENCES

(1) Golueke, C.G. and Oswald, W.J., "Harvesting and Processing Sewage-Grown Planktonic Algae," *Journ. W.P.C.F.*, 37, (4), 471-499, 1965.
(2) Golueke, C.G., Oswald, W.J. and Gotaas, H.B., "Anaerobic Digestion of Algae," *Applied Microbiology*, 5, 47-55, 1957.
(3) Golueke, C.G. and Oswald, W.J., "Biological Conversion of Light Energy to the Chemical Energy of Methane," *Applied Microbiology*, 7, 219-227, 1959.
(4) Oswald, W.J. and Golueke, C.G., "Solar Power via a Botanical Process," *Mechanical Engineering*, 86, (2), 40-43, 1964.
(5) Pfeffer, J.T., "Anaerobic Processing of Organic Refuse," *Proceedings of the Bioconversion Energy Research Conference*, Massachusetts University, Amherst, June 1973.
(6) Dague, R.R., "Application of Digester Theory to Digester Control," *Jour. Water Poll. Control Fed.*, 40, 2021-33, 1968.

BIBLIOGRAPHY

The following reports used in the preparation of this book are available from:

        National Technical Information Service
        U.S. Department of Commerce
        5285 Port Royal Road
        Springfield, Virginia 22151

AD/A 002 212   *Fuel from Organic Matter: Possibilities for the State of California,* D.J. Dugas, Rand Corporation, October 1973.

N75-25292   *Energy Recovery from Solid Waste,* Vol. 2, C.J. Huang and C. Dalton, University of Houston, Houston, Texas, April 1975.

ORNL-5056   *Anaerobic Mechanisms for the Degradation of Cellulose,* A.L. Compère and W.L. Griffith, Oak Ridge National Laboratory. June 1975.

PB 216 556   *Controlling Factors in Methane Fermentation,* R.E. Speece and R.S. Engelbrecht, New Mexico State University, University Park, N.M., 1964.

PB 220 821   *Production of Methane from Refuse and Sewage Sludge,* G.E. Johnson, Department of the Interior, Washington, D.C., March 1973.

PB 231 149   *Proceedings of the Bioconversion Energy Research Conference,* held at Massachusetts University, Amherst, on 25-26 June 1973.

PB 231 176   *Reclamation of Energy from Organic Waste,* J.T. Pfeffer, University of Illinois, Urbana, March 1974.

PB 235 468   *Biological Conversion of Organic Refuse to Methane,* J.T. Pfeffer and J.C. Liebman, University of Illinois, Urbana, July 1974.

Bibliography

PB 238 068   *Fuel Gas Production from Solid Waste,* R.G. Kispert, L.C. Anderson, D.H. Walker, S.E. Sadek, and D.L. Wise, Dynatech R/D Company, July 31, 1974.

PB 238 103   *Technology for the Conversion of Solar Energy to Fuel Gas,* University of Pennsylvania, Philadelphia, January 1974.

PB 238 291   *Rum Distillery Slops Treatment by Anaerobic Contact Process,* T.G. Shea, E. Ramos, J. Rodriguez and G.H. Dorian, Environmental Protection Technology Series, July 1974.

PB 238 563   *Fuel Gas Production from Solid Waste,* C.L. Cooney, E.E. Lindsey, R.S. Kirk and S. Oyewole, Dynatech R/D Company, July 31, 1974.

PB 240 113   *Animal Waste Conversion Systems Based on Thermal Discharge,* L. Boersma, E.W.R. Barlow, J.R. Miner and H.K. Phinney, Oregon State University, September 1974.

PB 240 768   *Urban Trash Methanation—Background for a Proof-of-Concept Experiment,* C. Bisselle, M. Kornreich, M. Scholl, and P. Spewak, MITRE Technical Report, February 1975.

PB 241 055   *Technology for the Conversion of Solar Energy to Fuel Gas,* National Center for Energy Management and Power, Oct. 31, 1973.

PB 245 083   *Fuel Gas Production from Solid Waste,* R.G. Kispert, S.E. Sadek, L.C. Anderson et al, Dynatech R/D Company, Jan. 31, 1975.

The following U.S. patent used in the preparation of this book is available from:

                Commissioner of Patents and Trademarks
                Washington, D.C. 20231

U.S. Patent 3,640,846; G.E. Johnson; February 8, 1972; assigned to the U.S. Secretary of the Interior.

HOW TO SAVE ENERGY AND CUT COSTS IN EXISTING INDUSTRIAL AND COMMERCIAL BUILDINGS 1976

An Energy Conservation Manual

by Fred S. Dubin, Harold L. Mindell
and Selwyn Bloome

Energy Technology Review No. 10

This manual offers guidelines for an organized approach toward conserving energy through more efficient utilization and the concomitant reduction of losses and waste.

The current tight supply of fuels and energy is unprecedented in the U.S.A. and other countries, and this situation is expected to continue for many years. Never before has there been as pressing a need for the efficient use of fuels and energy in all forms.

Most of the energy savings will result from planned systematic identification of, and action on, conservation opportunities.

Part I of this manual is directed primarily to owners, occupants, and operators of buildings. It identifies a wide range of opportunities and options to save energy and operating costs through proper operation and maintenance. It also includes minor modifications to the building and mechanical and electrical systems which can be carried out promptly with little, if any, investment costs.

Part II is intended for engineers, architects, and skilled building operators who are responsible for analyzing, devising, and implementing comprehensive energy conservation programs. Such programs involve additional and more complex measures than those in **Part I**. The investment is usually recovered through demonstrably lower operating expenses and much greater energy savings.

A partial and much condensed table of contents follows here:

PART I

1. **PRINCIPLES OF ENERGY CONSERVATION**
2. **MAJOR OPPORTUNITIES**
   Heating + Insulation
   Cooling + Insulation
   Lighting
   Hot Water
3. **ENERGY LOADS**
   Building Load
   Distribution Loads
   Equipment Efficiencies
   Building Profile vs. Energy Consumption
   Detailed Conservation Opportunities
4. **HEATING & VENTILATING**
   Primary Energy Conversion Equipment
   Building & Distribution Loads
   Energy Reducing Opportunities
5. **HOT WATER SAVINGS**
6. **COOLING & VENTILATING**
   Reducing Energy Consumption used for Cooling
7. **DISTRIBUTION & HVAC SYSTEMS**
8. **COMMERCIAL REFRIGERATION SYSTEMS**
9. **LIGHTING & HEAT FROM LAMPS**
   Fluorescents Turnoff vs. Replacement Cost & Labor
10. **POWER FOR MACHINERY**
APPENDIXES: COST VS. CONSERVATION

PART II

11. **ENERGY MANAGEMENT TEAMS**
12. **MECHANICAL SYSTEMS BACKGROUND**
13. **HEATING AND VENTILATION**
    Continuous Temperature Control
    Reduce Resistance to Air Flow
    Boiler & Burner Selection
14. **HOT WATER SYSTEMS**
15. **COOLING & VENTILATION**
    Condensers vs. Towers
    Losses Through Floors
    Air Shafts & Fenestration
16. **COMMERCIAL REFRIGERATION**
17. **HVAC & HEAT RECLAMATION**
18. **HEAT RECLAMATION EQUIPMENT**
19. **LIGHTING**
    Wattage Reduction
    Peak Load Reduction
    Non-Uniform Lighting
20. **POWER FOR MACHINERY**
    Correct Power Factors
    Exchange of Oversized Motors and Transformers
21. **CENTRAL CONTROL SYSTEMS**
    480/240 Volts Preferred to 120 V
22. **ALTERNATIVE ENERGY SOURCES**
    Solar Energy
    Methane Gas
23. **ECONOMIC ANALYSES & COSTS**
24. **APPENDIXES & REFERENCES**

Much of the technology required to achieve energy savings is already available. Current research is providing refinements and evaluating new techniques that can help to curb the waste inherent in yesteryear's designs. The principal need is to get the available technology, described here, into widespread use.

ISBN 0-8155-0638-4

725 pages

HYDROGEN TECHNOLOGY FOR ENERGY 1976

by David A. Mathis

Energy Technology Review No. 9

Hydrogen is attractive as a fuel because it is abundant, relatively inexpensive and ecologically clean. When hydrogen is burned in air, it forms water vapor only, there are no solid combustion residues and no soot particles or noxious gases to contaminate the atmosphere.

The use of hydrogen as a universal fuel necessitates development of methods for storing, handling, and transferring, and these are the main subjects treated in this book. Future volumes in this series will be reserved for the technology of the myriad of hydrogen production schemes now under consideration viz. electrolysis of seawater by ocean-derived thermal energy, by solar cells, certain algae, by nuclear powered direct thermochemical conversion, or from municipal waste, etc.

The first chapter describes the hydrogen economy and suggests how it can be integrated into the USA energy system. The next three chapters are concerned with handling the various forms of hydrogen: gas, liquid, and solid (in the form of metal hydrides). The fifth chapter describes some of the work which has been done or is under way in using hydrogen as a fuel or in an energy storage system. Another chapter delves into safety and the political, socioeconomic, and environmental implications of a hydrogen economy. The final chapter provides a list of hydrogen technology experts and includes a brief description of each individual's expertise in the various aspects of hydrogen technology.

A partial and condensed table of contents follows here.

1. THE HYDROGEN ECONOMY
Primary Energy Sources
Hydrogen Energy Conversion
Hydrogen Production
Transportation and Storage
Utilization of Hydrogen
Advantages and Disadvantages

2. GASEOUS HYDROGEN
Comparison of Storage Methods
Hydrogen Handling Systems
Economics of Transmission
Pipeline Systems
Embrittlement & Compatibility
High Pressure Systems
Pressure Vessels & Pipelines
Compressor Requirements
Regenerative Compressors
Cost of Compression Equipment

3. LIQUID HYDROGEN
Technology of Liquefaction
Costs (Capital & Running)
Recovery of Liquefaction Energy
Para-Hydrogen
Storage and Transfer
Storage Dewars
Losses & Fire Hazards
Pumping Problems
The "Energy Pipe"
Peak Shaving
Energy Transmission Costs for
  Near-Urban Environments

4. SOLID HYDRIDES
Hydride Storage
Heat Transfer
Deterioration
Metal Hydride Suitability
Iron-Titanium-Manganese Alloy
Fixed Bed Metal Hydride Storage
Transmission of H_2 from Metal
  Hydride "Compressors"

5. FUEL & ENERGY STORAGE & USE
Water-Modified Aphodid Burner
Hydrogen Fueled Engines
Energy Storage for Utilities
H_2-Fuel for Transportation
Vehicle Fuel Storage
Metal Hydrides for Vehicles
Hydride-Dehydride Power
  System for Refrigeration

6. NONTECHNICAL ASPECTS OF A HYDROGEN ECONOMY
Safety Implications
Energy Law
Regulatory Law
Environmental Law and Implications
Unlikely NO_x Production
  at High Temperatures
Possible Production of Reject Heat
Underground Pipeline Transmission
International Implications

7. HYDROGEN TECHNOLOGY EXPERTS
A Listing of 263 Technological Experts,
  their Affiliations and Addresses,
  arranged by their Specialties

BIBLIOGRAPHY

ISBN 0-8155-0629-5

285 pages

THERMAL ENERGY FROM THE SEA 1975

by Arthur W. Hagen

Energy Technology Review No. 8
Ocean Technology Review No. 5

Recent advances in heat transfer research and thermal power plant design suggest that sea thermal power can now be made competitive with more conventional generating methods.

Utilization of ocean thermal gradient systems appears relatively attractive from many points of view. The ocean acts as a large solar energy heat reservoir which reduces energy storage requirements and permits the system to be operated the year round, 24 hours per day.

The purpose of this book is to provide a condensed data base to aid in proof-of-concept experiments and continued R & D to prove the technical feasibility and economic viability of generating either electricity or hydrogen by harnessing the temperature gradients in the sea.

The optimum locations for such generating plants appear to be in latitudes from 23°N to 23°S. Transmission and storage problems may limit the amount of power delivered by ocean-based plants. Construction of power production facilities must consider design problems associated with the hazards of marine environment.

Reports of proposed solutions to these problems are also contained in this book which is based on government-sponsored studies by engineering firms and university research teams. A partial and condensed table of contents follows:

1. **AN OVERVIEW OF SEA THERMAL POWER**
   Introductory Material and NASA Reports
   Site Analysis
   Cost
   Systems Analysis

2. **OPEN CYCLE THERMAL GRADIENT OCEANIC POWER PLANT**
   Introductory Material
     NASA and PB Reports
   Preliminary Design Investigation
   Overall System Design
   Spray Evaporator Design
   Turbine Design
   Deaeration Losses
   Condenser Design
   Cold Water Pump Considerations
   Fresh Water Production
   Feasibility Study of a
     100 Megawatt Plant

3. **CARNEGIE MELLON UNIVERSITY DESIGN**
   Introductory Material and PB Reports
   Design Overview
   Water Circulation
   Heat Transfer Fluid Circulation
   Seawater Inlet-Outlet Hydrodynamics
   Boiler Technology Considerations
   Cold Water Pipe vs. Ammonia Pipe
   Vertical Tube Heat Exchangers
   Antifouling Costs
   SSPP (**S**olar **S**ea **P**ower **P**lant) Topology

4. **UNIVERSITY OF MASSACHUSETTS DESIGN**
   Introductory Material and PB Reports
   Overall Design Concept (Mark I)
   Cold Water Supply & Suction Pipe
   Anchor & Mooring Systems
   Energy Umbilical
   Control of Biofouling
   Heat Exchanger Design (Mark I)
   Variations and Improvements (Mark II)
   Evaluation of Turbomachinery
   Technical & Economic Feasibility Studies

5. **COMPARATIVE STUDY OF CLOSED CYCLE TECHNOLOGIES**
   NASA Reports
   CMU vs. UMASS Designs
   Cost Evaluations

6. **ADDITIONAL TECHNOLOGIES**
   Hydrogen Utilization
   Hydrogen Energy
   NITINOL (Ni-Ti-Naval Ordnance Lab.)
   #55 Alloy Utilization
   Further SSPP Proposals
   OTEC Proposal (Ocean Thermal
     Energy Conversion Proposal)
   The Danish System
   A Japanese System

ISBN 0-8155-0597-3   150 pages

SOLAR ENERGY FOR HEATING AND COOLING OF BUILDINGS 1975

by Arthur R. Patton

Energy Technology Review No. 7

Solar energy can be used for indirect heating purposes in many ways. The information in this book has been limited to so-called low temperature solar thermal processes. Designs requiring photocells or other thermoelectric generators and lenses or reflecting mirrors plus tracking equipment have been excluded.

Low temperatures are the easiest to obtain, and the necessary collectors are fairly simple in construction. A black surface is used to absorb the sun's rays, this surface is usually covered with glass and the collector is insulated on the back and sides against heat loss. Water or some other heat transfer fluid is passed through the collector and can reach temperatures from 60°C (140°F) to about 95°C (203°F). The thermal energy is then stored in a heat storage system (perhaps based on the latent heat of fusion of selected salts). Coupled to the heat storage system are heating loops to furnish heat by convection and to operate an air conditioning system. In most temperate zones an auxiliary heater, operated with conventional fuels, must also be connected.

Large scale applications designed for schools and similar building are beginning to appear or are in the planning stage. This book describes in detail several large scale feasibility studies with designs suitable for institutions and industrial plants.

Descriptions are based on studies conducted by industrial or engineering firms or university research teams under the auspices of various government agencies. A partial and condensed table of contents follows here.

1. SYSTEM COMPONENTS
 Collectors—Flat Plate Construction
 Loss Control Concepts
 Internal Heat Transfer
 Heat Storage Designs
 Solar Powered Air Conditioning

2. HISTORY OF EXPERIMENTAL SYSTEMS
 System Configurations
 Air as the Working Fluid
 Water as the Working Fluid
 Integral Collector-Storage Configurations
 Other Configurations

3. DETAILED DESCRIPTIONS AND LOCATIONS OF EXPERIMENTAL SYSTEMS
 Heating Systems
 Cooling and Refrigeration Systems

4. SIMULATED SYSTEMS
 University of Wisconsin Study
 Albuquerque Model
 System Models
 Systems Evaluation and Cost Analyses
 Marshall Space Flight Center
 Background
 Description of Test System
 Model Techniques
 Simulation Results
 University of Pennsylvania Study

5. FEASIBILITY STUDIES FOR LARGE SCALE INSTALLATIONS
 National Science Foundation Study
 Overview
 Evaluation of Present
  Status of Technology
 Implementation Studies
 MITRE Study
 General Assessments
 Problem Areas
 Improvement Studies
 Economic Viability Calculations

6. GENERAL ELECTRIC STUDY
 Scenario Projections
 Systems Descriptions
 Technical Evaluations
 Capture Potential and
  Projections of Market Penetration
 POCE—Proof of Concept Experiments

7. WESTINGHOUSE STUDY
 Energy Consumption
 Operational Requirements
 Capture Potential and Master Plan

8. TRW STUDY
 Climatic Region Classifications
 Reference System Designs

9. COLORADO STATE UNIVERSITY INTEGRATED SYSTEM DESIGN
 Project Objectives and Data Handling
 Desirable Building Designs
 Lithium Bromide Absorption Cooling Unit

10. OTHER RESEARCH

11. AVAILABLE HARDWARE

12. NAMES & ADDRESSES FOR FURTHER INFORMATION

13. REFERENCES

ISBN 0-8155-0579-5

WIND POWER 1975

by Daniel M. Simmons

Energy Technology Review No. 6

Wind is a free, clean, inexhaustible source of energy. In previous centuries, when dependence on time and deadlines for production and shipments were not as important as they are now, wind served mankind well, despite becalmed and stormy intervals. Wind moved sailing ships of all sizes and drove windmills to grind seeds and other materials, saw wood and to pump water. In later years it was used to produce some electrical current of the DC type to charge storage batteries.

There are three phases to harnessing the power of the wind. The first is the selection of a suitably windy site. This necessitates studies of wind behavior in possible locales, including ocean sites. Secondly a suitably designed windmill and generator to convert the mechanical energy or to convert seawater to hydrogen and oxygen. The third and rather fundamental difficulty is the available choice of storage batteries or other storage means, so that energy produced at peak times can be used when the wind is not blowing.

The various phases for developing wind power are considered in the first three chapters of this book. The next three describe wind power developments in the U.S., Canada, U.S.S.R., Germany, Denmark, France, U.K., Sweden and other countries in Asia and Africa. The last chapter describes commercially available apparatus.

This Energy Technology Review is based primarily on international studies conducted by industrial and engineering firms or university research teams under the auspices of various governments and governmental agencies.

A partial and condensed table of contents follows here.

1. WIND BEHAVIOR AND SITE SELECTION
Influence of Height Above Ground
Wind Surveys and Site Selection
Anemometers and Wind Speed Recorders
Isovent Cards
Use of Metereological Data
Wind Studies in Different Countries

2. WIND MACHINE DESIGN
Major Problems of Wind Power
Tower Height & Design
Complete Systems
Wind Turbines
Small Wind Electric Plant with
   Permanent Magnet Generator
Sailwing Windmills
Vertical Axis Wind Rotors
Diffuser-Augmented Wind Turbine
The J. D. Madaras Project

3. WIND CONVERSION AND STORAGE SYSTEMS
Variable Speed Input vs.
   Constant Current Output
Asynchronous AC/DC/AC Converter
Seawater Electroylsis to
   H_2 and O_2 and Their Storage
Compressed Air Manufacture
Batteries
Superflywheels
Alternate Systems
Flow Charts for Alternate Systems

4. WIND POWER DEVELOPMENT IN THE U.S. AND CANADA
Small Wind Power Systems
An Economic Study
Large Scale Applications
Ocean-Based Wind Machines
Preliminary Design of a 100 KW
   Wind Turbine Generator
Proposed Regions

5. U.S.S.R. DEVELOPMENTS
Large Scale Wind Generator
Agricultural Usage
Wind Machine Layouts
Water Lifting Machines
Other Wind Powered Machines
Wind Machines for Making Electricity
Typically Russian Designs

6. WIND POWER DEVELOPMENTS IN OTHER COUNTRIES
Netherlands and adjacent Low Countries
West Germany & Denmark
France & Great Britain
Sweden
India
Israel
Algeria
Curaçao

7. COMMERCIALLY AVAILABLE WIND MACHINES
Aermotor
Automatic Power, Inc.
Dempster Industries, Inc.
Edmund Scientific Co.
Environmental Energies, Inc.
Heller-Aller Co., Inc.
Solar Wind Co.
Winco
Windworks

ISBN 0-8155-0575-2

300 pages

GEOTHERMAL ENERGY 1975

by Edward R. Berman

Energy Technology Review No. 4

This book describes in detail the nature of the geothermal resources, their extent, and the currently available technology by which these natural sources of energy can be exploited and utilized to the greatest advantage. There is little pollution resulting from the use of geothermal energy, and with a little care all disturbance of natural ecological systems can be avoided.

Earth heat can be used most practically where hot volcanic rocks are comparatively near the surface, and emerging circulating ground waters act as heat collectors, either by producing steam or by serving as heat transfer media. Suitable sites have been discovered in the Continental U.S. and Hawaii, the U.S.S.R., Japan, New Zealand, Iceland, and Italy. In view of the rise in the price of oil and the intensive search for new sources of energy, geothermal power appears to be in for a period of rapid development.

This Energy Technology Review is based on international studies conducted by industrial and engineering firms or university research teams under the auspices of various governments and governmental agencies.

A partial and condensed table of contents follows here.

1. U.S. RESEARCH & EXPLORATION TECHNIQUES
Historical Background
Recent Explorations
U.S. Agencies Involved
World Survey of Major Geothermal Installations (except USSR)

2. RESEARCH AND PRACTICE IN THE USSR
Soviet Geothermal Electric Power Generation
Paratunka Geothermal Electric Power Station (using Freon®)
Pauzhetka Power Station
Bol'she-Bannaya Station
Makhachkala Station
Additional Sites
Non-Electric Uses of Thermal Waters
Equipment and Instruments

3. DRY GEOTHERMAL RESERVOIRS
Conventional Drilling
Ultra-Deep Drilling

Limitations of Current Technology
Areas of Further Research
Novel Techniques
A Proposed Commercial System
Man-Induced Fracturing of Hot Rocks
Artificial Geothermal Reservoirs

4. THE PLOWSHARE CONCEPT
Geothermal Heat Extraction Using Nuclear Explosives
Site Guidelines & Geology
Heat Source Developments
System Thermodynamics
Seismic Design Considerations

5. EXPERIMENTAL STUDIES
Scale Formation in Simulated Geothermal Brine
Wairakei (New Zealand) Well Analysis
Aquifer Chemistry in Geothermal Systems
Plowshare Geothermal Steam Chemistry

6. GEOTHERMAL RESOURCES OF CALIFORNIA
The Geysers
Imperial Valley
Mono Lake
Desalination of Geothermal Water
Supplemental Water for Reinjection
Land Subsidence and Cave-Ins
System Synthesis and Costs

7. ENERGY RECOVERY FROM NATURAL HOT BRINE
Chemical Composition
Fluid Pressures
Turbine Systems
Corrosion and Scale
Sperry Rand Corp. Method

8. FEASIBILITY STUDIES FOR SPECIFIC U.S. AREAS
Texas
  Port Mansfield Site
California
  Point Mugu Site
  Twenty-Nine Palms Site
  China Lake Site
Idaho
  Raft River Valley Site

9. PROPOSED RESEARCH

ISBN 0-8155-0563-9

336 pages